TWO-WEEK WAIT

Luke C. Jackson is a teacher and the author
of novels, games, and films. Kelly Jackson is a
teacher and educational writer. They began their
own IVF journey in 2011 and are now parents
of two daughters.

Mara Wild lives and works in Hamburg, Germany,
as a freelance illustrator and animator, mostly doing
multimedia artwork. She met Luke while studying
abroad in Melbourne, taking one of his classes.

Two-Week Wait is their first graphic novel.

TWO-WEEK WAIT

AN I.V.F. STORY

LUKE C. JACKSON   KELLY JACKSON   MARA WILD

SCRIBE

Melbourne • London

Scribe Publications
18–20 Edward St, Brunswick, Victoria 3056, Australia
2 John Street, Clerkenwell, London, WC1N 2ES, United Kingdom
3754 Pleasant Ave, Suite 100, Minneapolis, Minnesota 55409, USA

First published by Scribe 2021

Printed and bound in Singapore by C.O.S. Printers Pte Ltd

Scribe Publications is committed to the sustainable use of natural
resources and the use of paper products made responsibly from
those resources.

9781925713824 (Australian edition)
9781913348649 (UK edition)
9781950354634 (US edition)
9781925938838 (ebook)

Catalogue records for this book are available from the National
Library of Australia and the British Library.

scribepublications.com.au
scribepublications.co.uk
scribepublications.com

To everybody who helped us through our IVF journey.

I.C.J and K.J.

CHAPTER 1:
TRYING

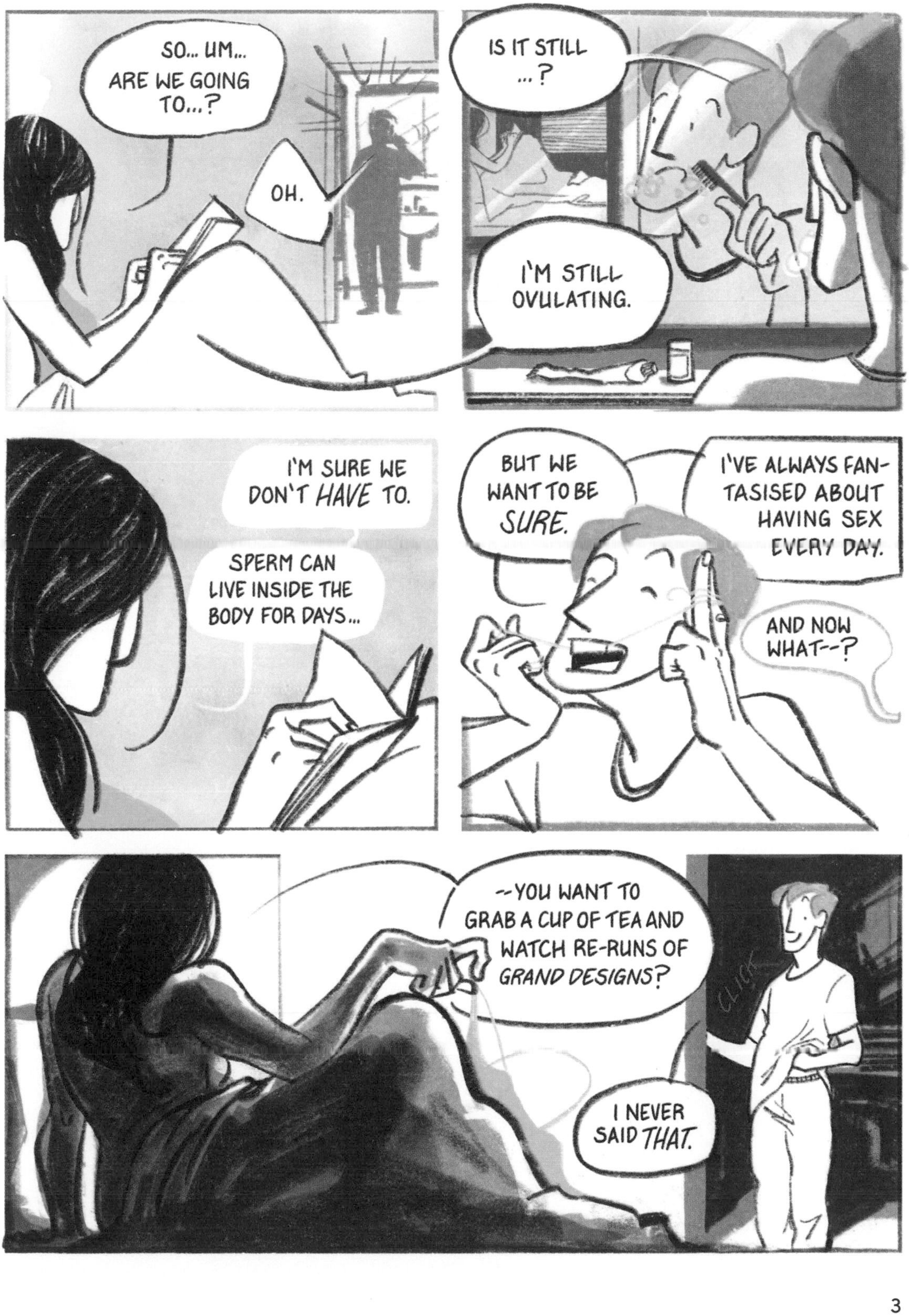
SO... UM...
ARE WE GOING TO...?
OH.
IS IT STILL ...?
I'M STILL OVULATING.
I'M SURE WE DON'T HAVE TO.
SPERM CAN LIVE INSIDE THE BODY FOR DAYS...
BUT WE WANT TO BE SURE.
I'VE ALWAYS FANTASISED ABOUT HAVING SEX EVERY DAY.
AND NOW WHAT--?
--YOU WANT TO GRAB A CUP OF TEA AND WATCH RE-RUNS OF GRAND DESIGNS?
CLICK
I NEVER SAID THAT.

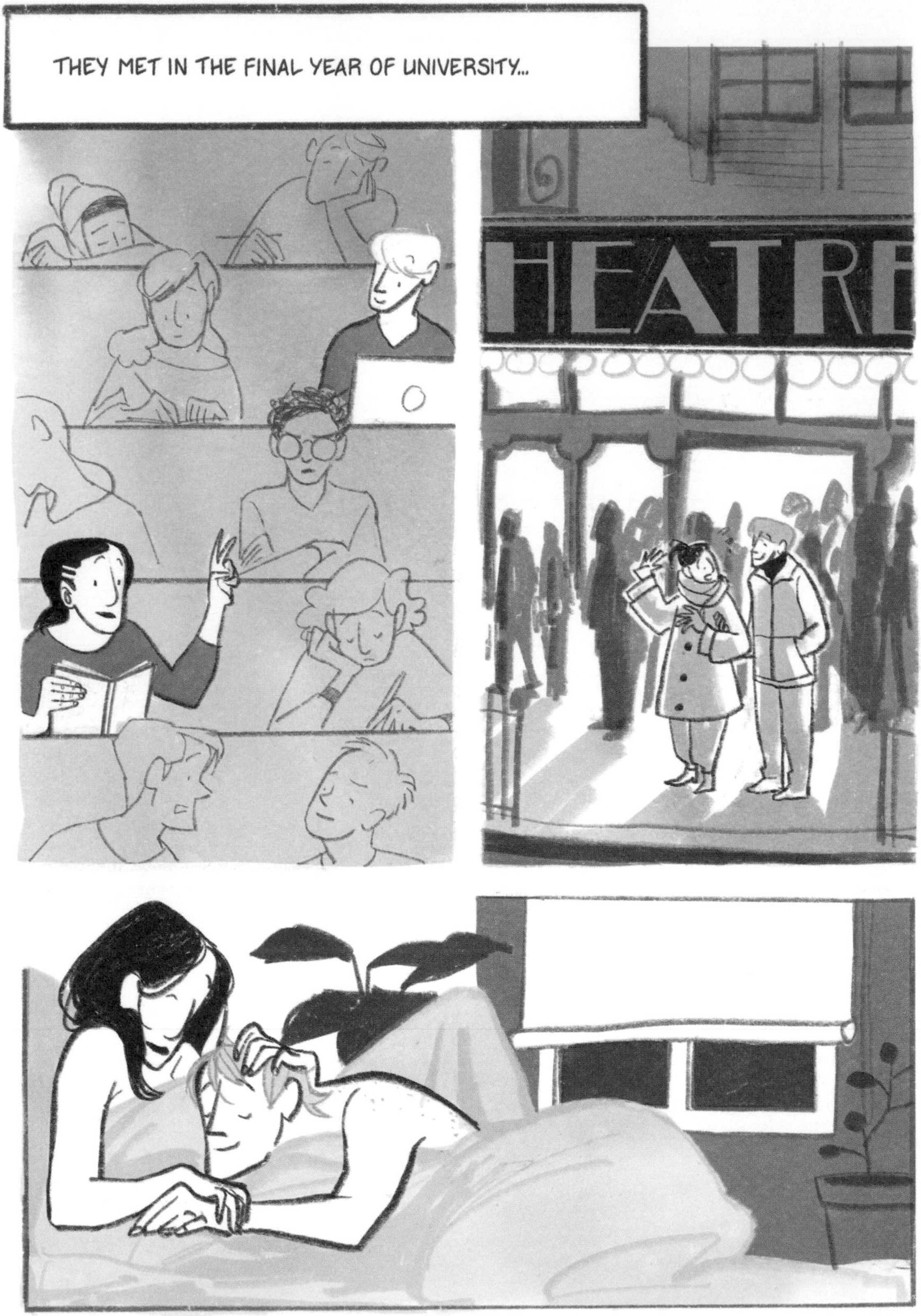
THEY MET IN THE FINAL YEAR OF UNIVERSITY...
HEATRE

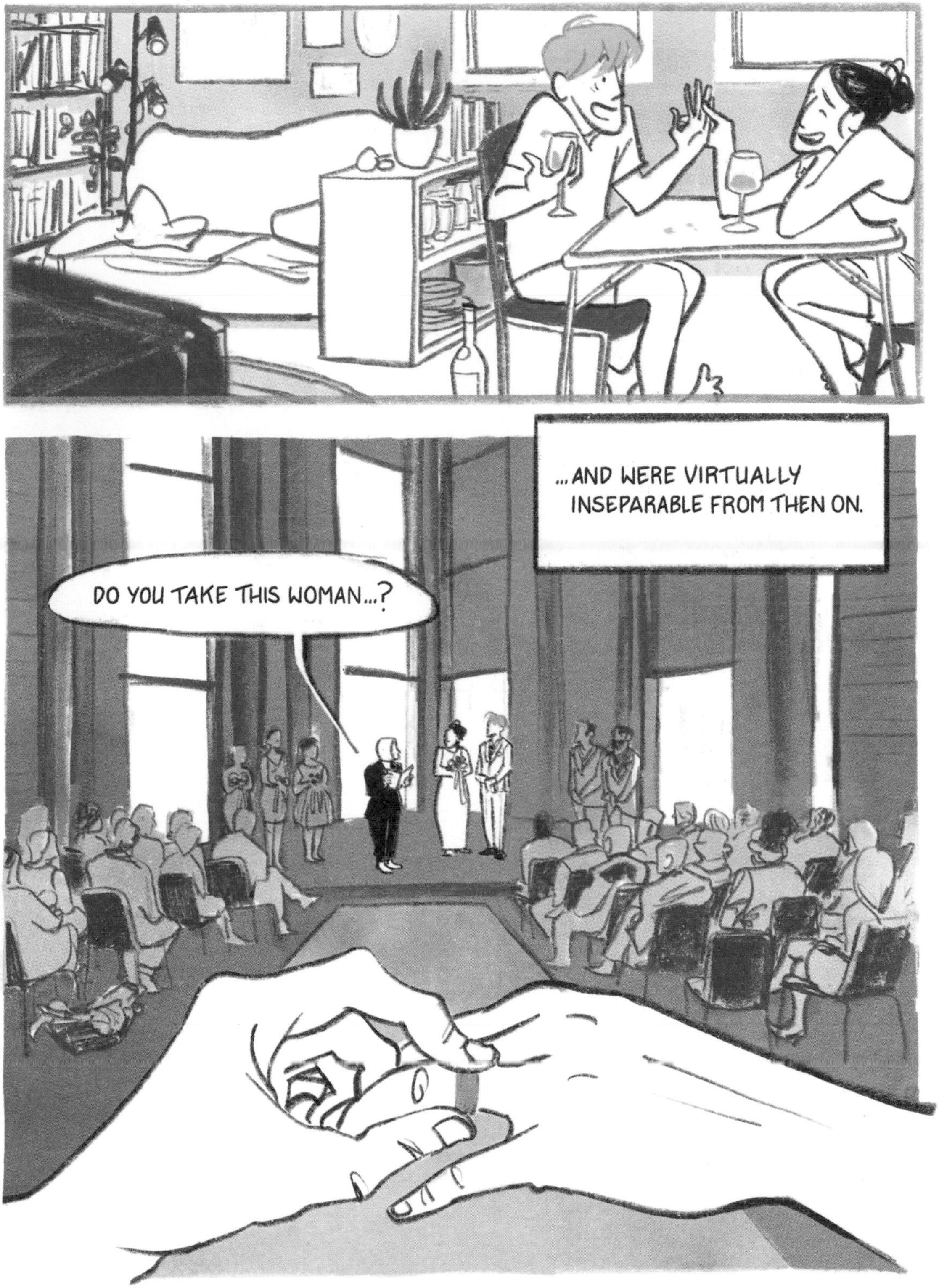
...AND WERE VIRTUALLY INSEPARABLE FROM THEN ON.
DO YOU TAKE THIS WOMAN...?

FOR A WHILE, IT FELT LIKE THEY HAD ALL THE TIME IN THE WORLD...

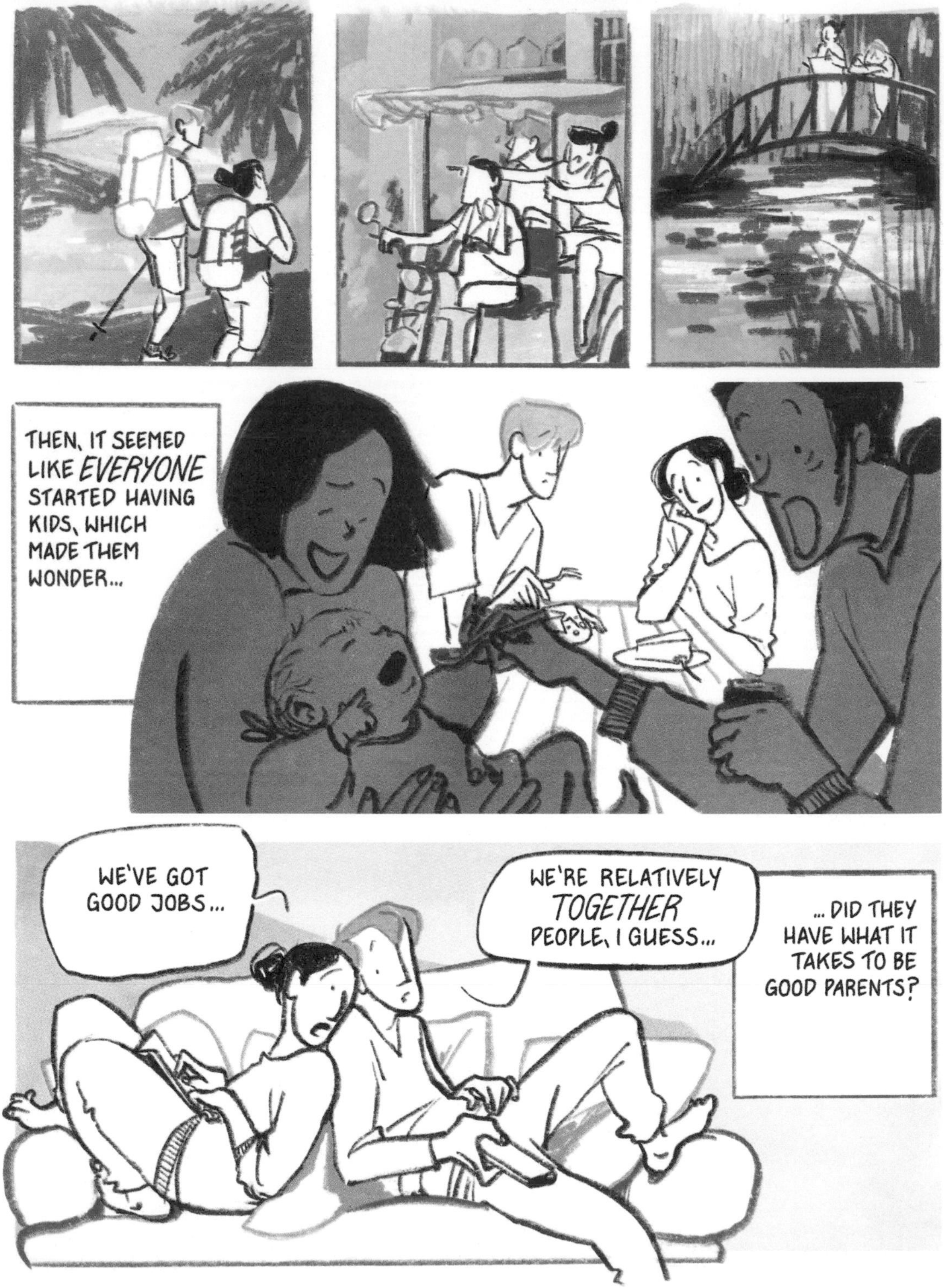
THEN, IT SEEMED LIKE *EVERYONE* STARTED HAVING KIDS, WHICH MADE THEM WONDER...
WE'VE GOT GOOD JOBS...
WE'RE RELATIVELY *TOGETHER* PEOPLE, I GUESS...
... DID THEY HAVE WHAT IT TAKES TO BE GOOD PARENTS?

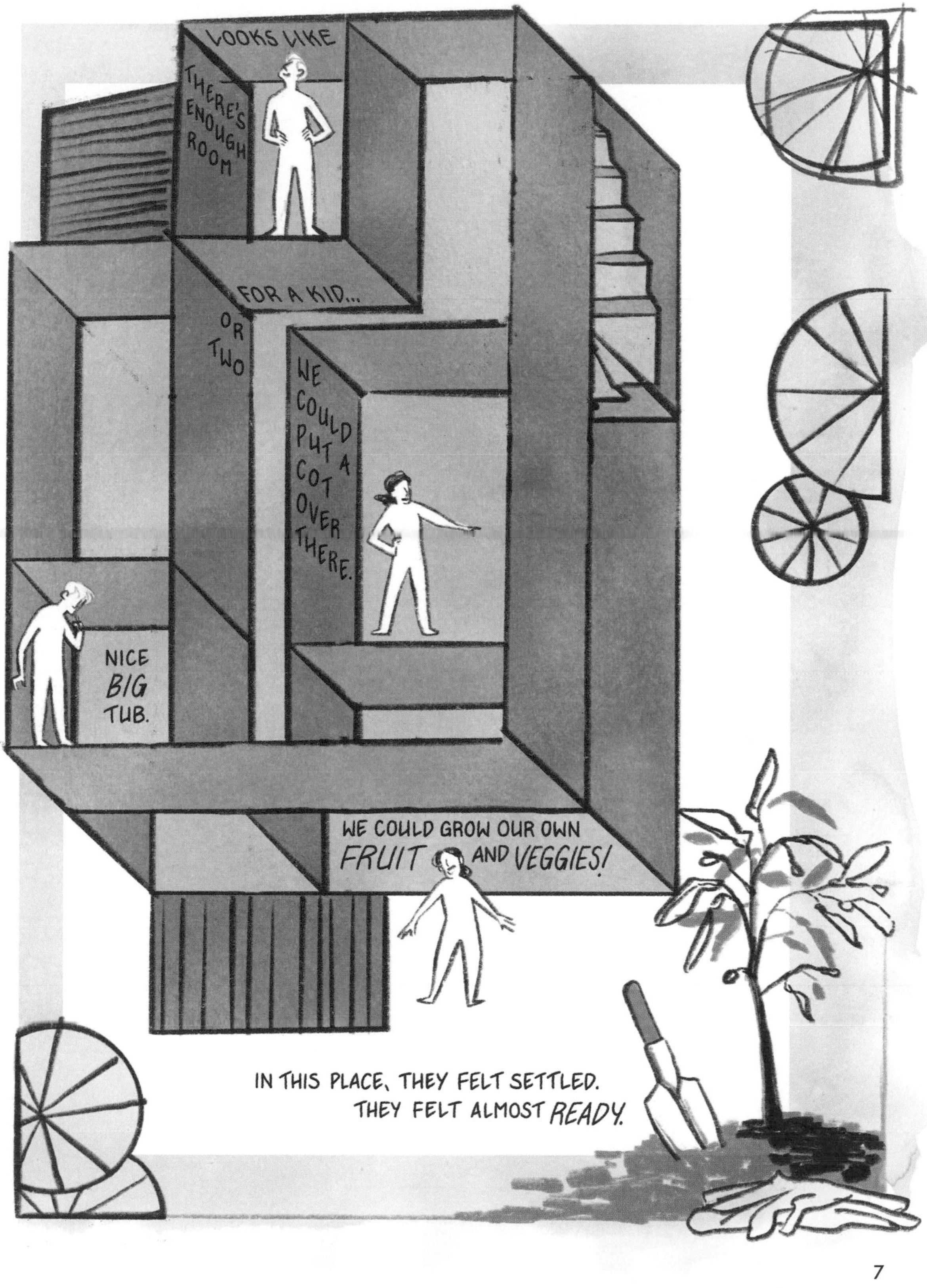
LOOKS LIKE
THERE'S ENOUGH ROOM
FOR A KID...
OR TWO
WE COULD PUT A COT OVER THERE.
NICE BIG TUB.
WE COULD GROW OUR OWN FRUIT AND VEGGIES!
IN THIS PLACE, THEY FELT SETTLED.
THEY FELT ALMOST READY.

...BUT, A YEAR LATER, THEY WERE STILL *TRYING.*

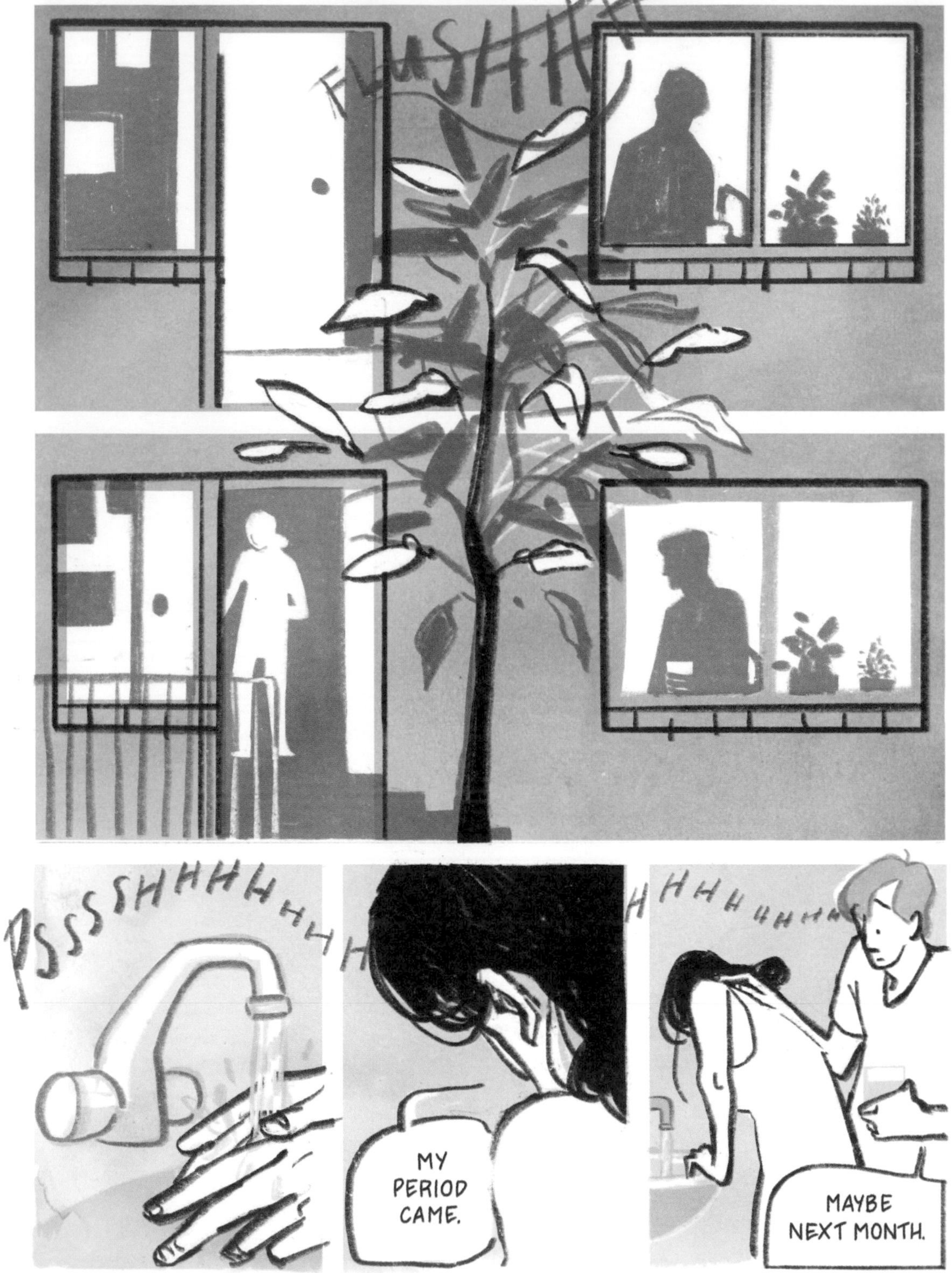

FLUSHHH
PSSSSHHHHHHHHH
HHHHHHHHH
MY PERIOD CAME.
MAYBE NEXT MONTH.

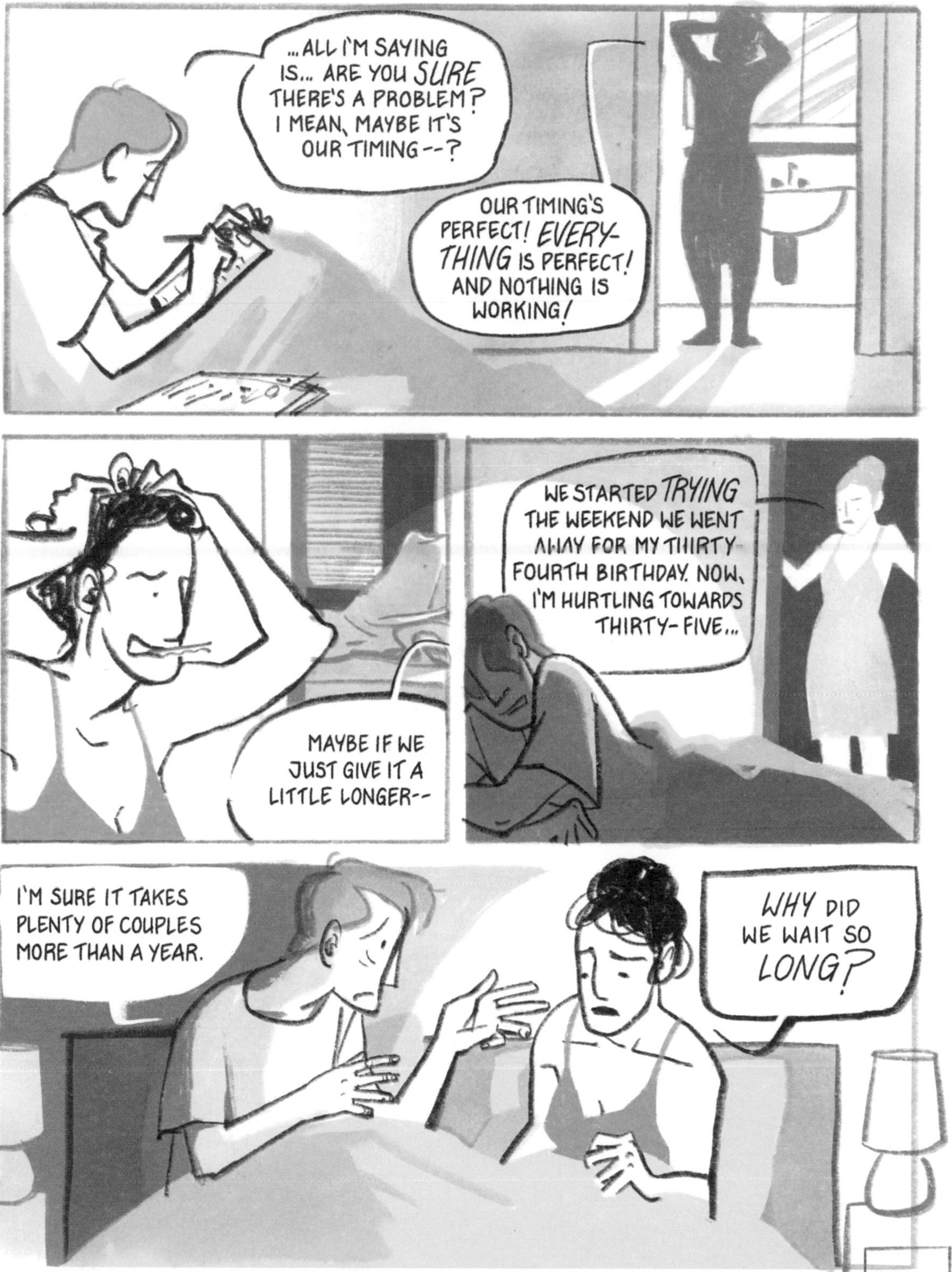
...ALL I'M SAYING IS... ARE YOU SURE THERE'S A PROBLEM? I MEAN, MAYBE IT'S OUR TIMING--?
OUR TIMING'S PERFECT! EVERY-THING IS PERFECT! AND NOTHING IS WORKING!
WE STARTED TRYING THE WEEKEND WE WENT AWAY FOR MY THIRTY FOURTH BIRTHDAY. NOW, I'M HURTLING TOWARDS THIRTY-FIVE...
MAYBE IF WE JUST GIVE IT A LITTLE LONGER--
I'M SURE IT TAKES PLENTY OF COUPLES MORE THAN A YEAR.
WHY DID WE WAIT SO LONG?

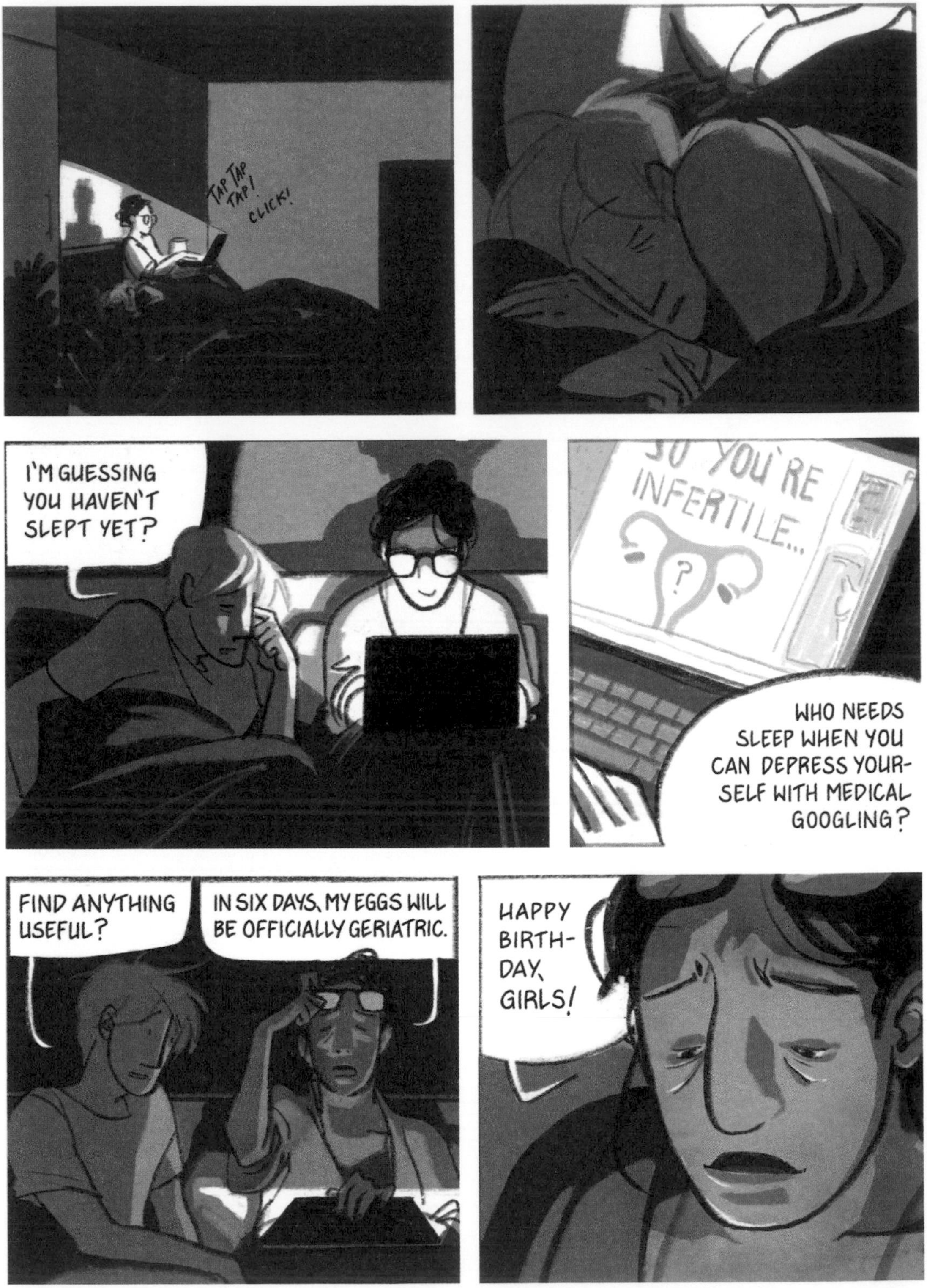
TAP TAP TAP!
CLICK!
I'M GUESSING YOU HAVEN'T SLEPT YET?
SO YOU'RE INFERTILE...
WHO NEEDS SLEEP WHEN YOU CAN DEPRESS YOURSELF WITH MEDICAL GOOGLING?
FIND ANYTHING USEFUL?
IN SIX DAYS, MY EGGS WILL BE OFFICIALLY GERIATRIC.
HAPPY BIRTHDAY, GIRLS!

I CALLED THE G.P.

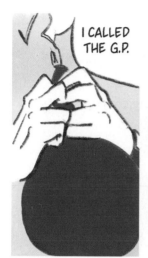

HE'S BOOKED ME IN FOR AN APPOINTMENT EARLY NEXT WEEK.

AND—

UM—

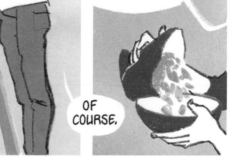

YOU'LL REMEMBER TO GET A REFERRAL FOR THE FERTILITY TESTING?

OF COURSE.

URG!

SORRY.

I FEEL LIKE ALL I'M DOING IS *NAGGING* YOU AT THE MOMENT.

WHY DID I THINK IT WAS A GOOD IDEA TO HAVE A PARTY RIGHT NOW?

BECAUSE YOU KNOW YOU SHOULD TAKE A NIGHT *OFF!*

YOU'RE RIGHT!

DING!

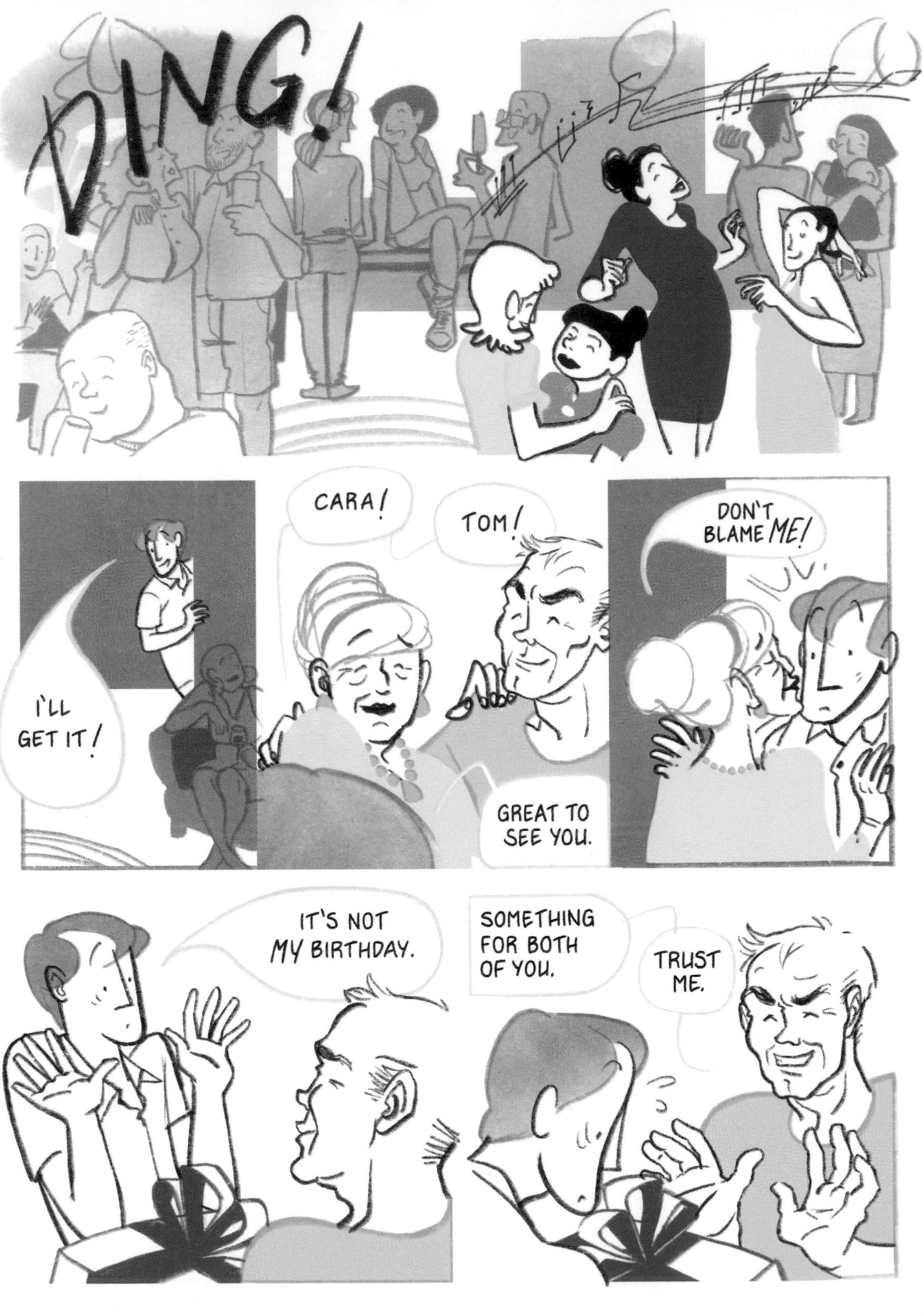
DING!
I'LL GET IT!
CARA!
TOM!
GREAT TO SEE YOU.
DON'T BLAME ME!
IT'S NOT MY BIRTHDAY.
SOMETHING FOR BOTH OF YOU.
TRUST ME.

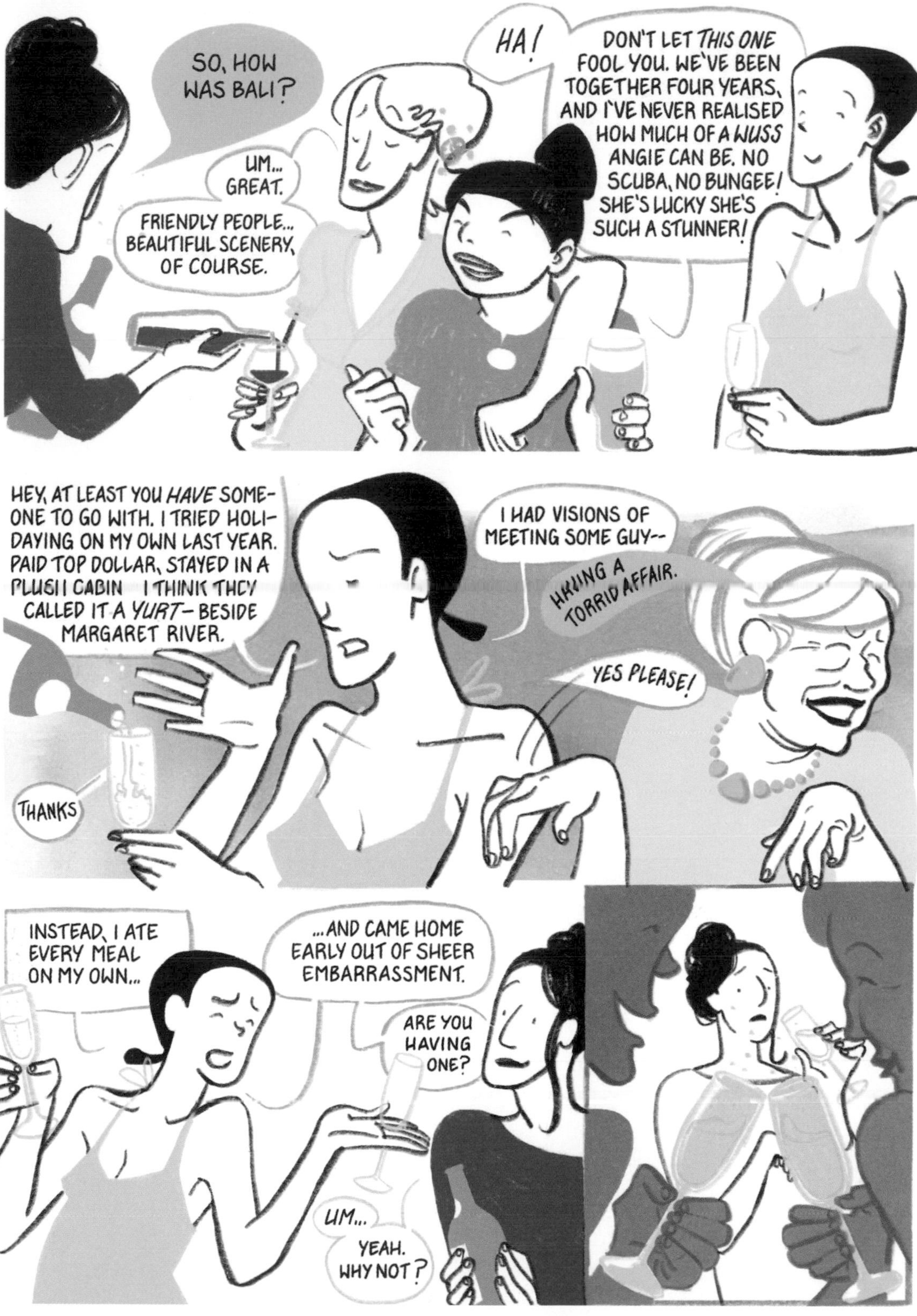
SO, HOW WAS BALI?
UM... GREAT.
FRIENDLY PEOPLE... BEAUTIFUL SCENERY, OF COURSE.
HA!
DON'T LET THIS ONE FOOL YOU. WE'VE BEEN TOGETHER FOUR YEARS, AND I'VE NEVER REALISED HOW MUCH OF A WUSS ANGIE CAN BE. NO SCUBA, NO BUNGEE! SHE'S LUCKY SHE'S SUCH A STUNNER!
HEY, AT LEAST YOU HAVE SOMEONE TO GO WITH. I TRIED HOLIDAYING ON MY OWN LAST YEAR. PAID TOP DOLLAR, STAYED IN A PLUSH CABIN—I THINK THEY CALLED IT A YURT—BESIDE MARGARET RIVER.
THANKS
I HAD VISIONS OF MEETING SOME GUY--
HAVING A TORRID AFFAIR.
YES PLEASE!
INSTEAD, I ATE EVERY MEAL ON MY OWN...
...AND CAME HOME EARLY OUT OF SHEER EMBARRASSMENT.
ARE YOU HAVING ONE?
UM...
YEAH. WHY NOT?

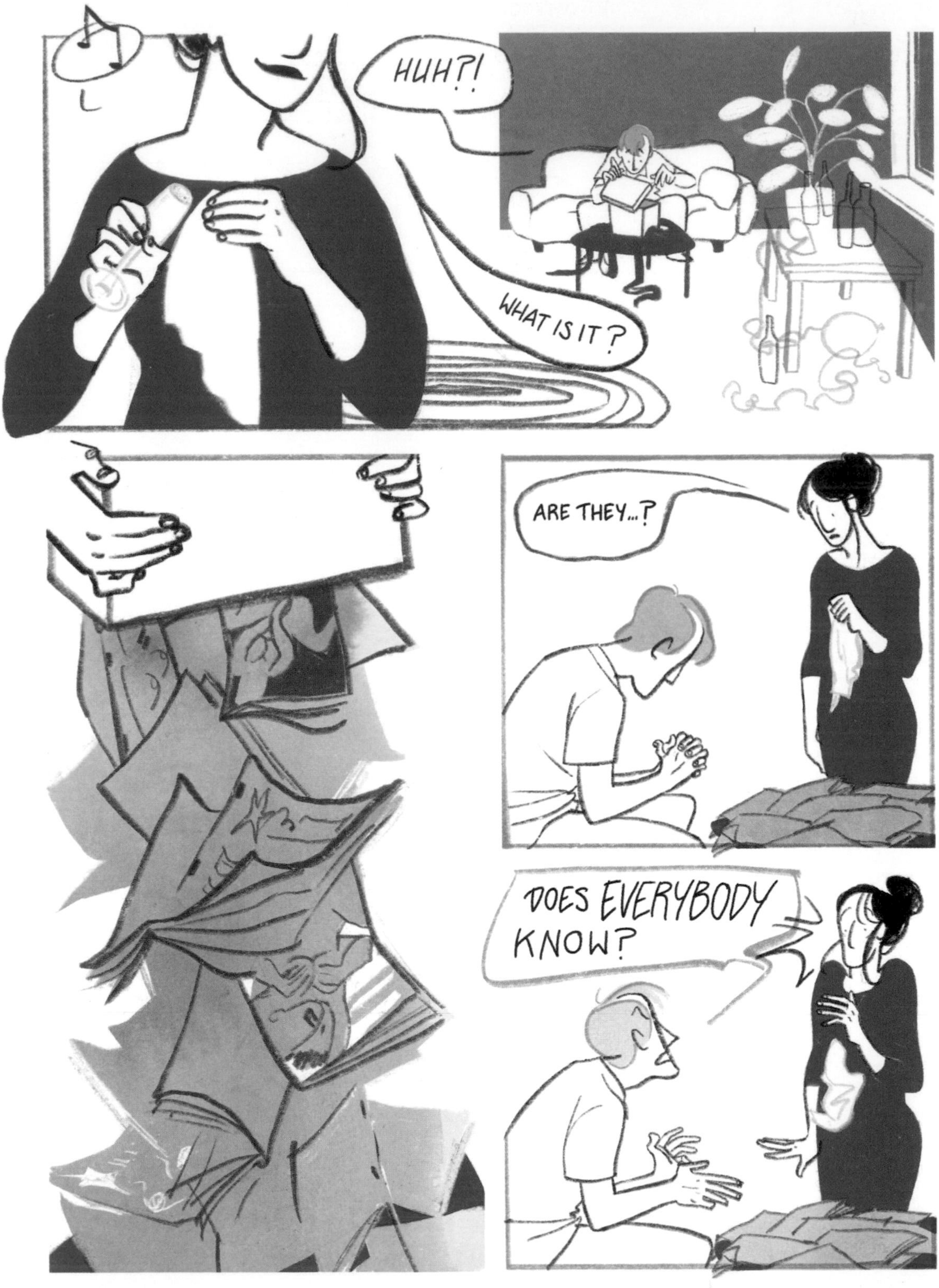
HUH?!
WHAT IS IT?
ARE THEY...?
DOES EVERYBODY KNOW?

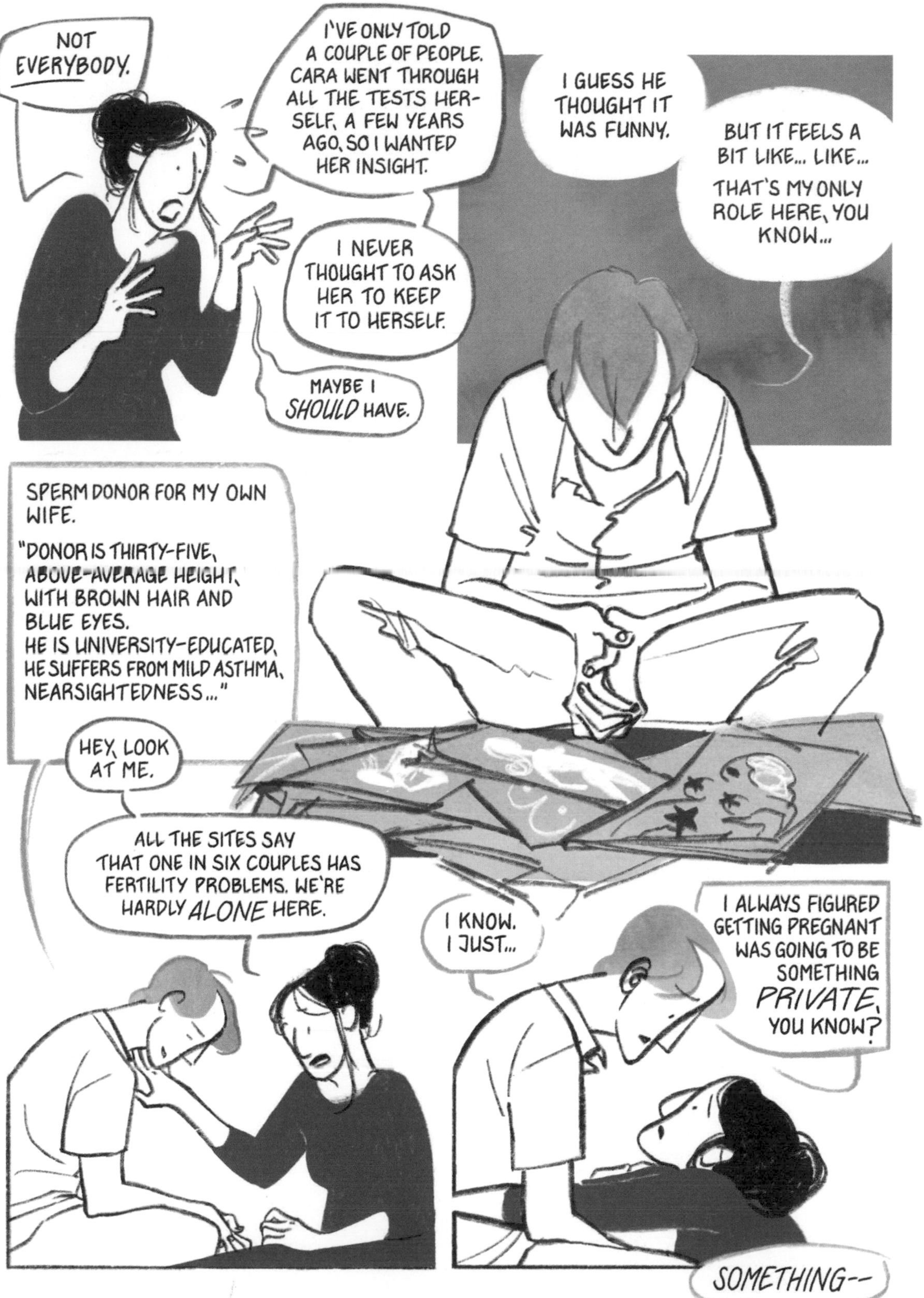
NOT EVERYBODY.
I'VE ONLY TOLD A COUPLE OF PEOPLE. CARA WENT THROUGH ALL THE TESTS HERSELF, A FEW YEARS AGO, SO I WANTED HER INSIGHT.
I NEVER THOUGHT TO ASK HER TO KEEP IT TO HERSELF.
MAYBE I SHOULD HAVE.
I GUESS HE THOUGHT IT WAS FUNNY.
BUT IT FEELS A BIT LIKE... LIKE...
THAT'S MY ONLY ROLE HERE, YOU KNOW...
SPERM DONOR FOR MY OWN WIFE.
"DONOR IS THIRTY-FIVE, ABOVE-AVERAGE HEIGHT, WITH BROWN HAIR AND BLUE EYES.
HE IS UNIVERSITY-EDUCATED, HE SUFFERS FROM MILD ASTHMA, NEARSIGHTEDNESS..."
HEY, LOOK AT ME.
ALL THE SITES SAY THAT ONE IN SIX COUPLES HAS FERTILITY PROBLEMS. WE'RE HARDLY ALONE HERE.
I KNOW. I JUST...
I ALWAYS FIGURED GETTING PREGNANT WAS GOING TO BE SOMETHING PRIVATE, YOU KNOW?
SOMETHING--

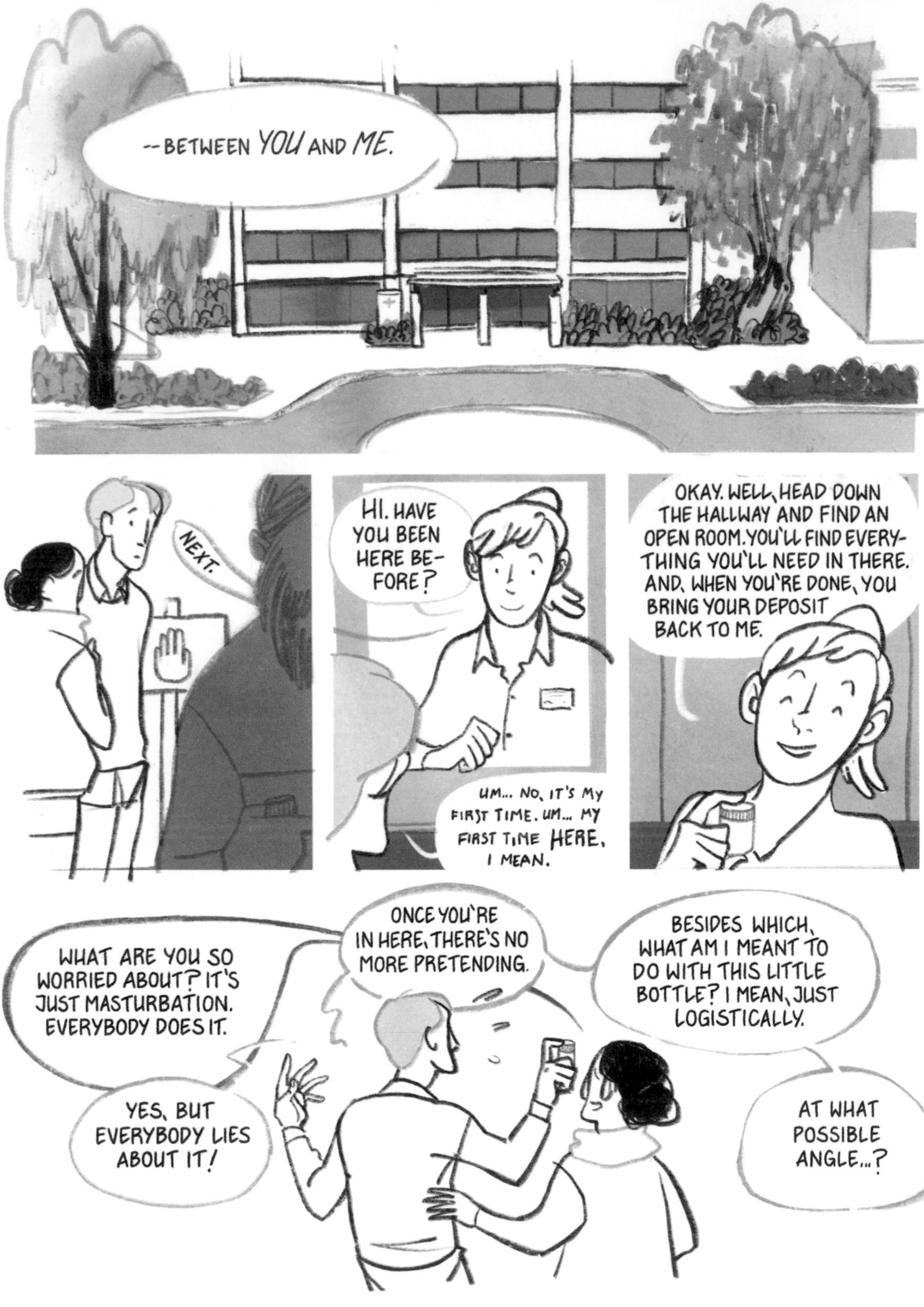
--BETWEEN YOU AND ME.
NEXT.
HI. HAVE YOU BEEN HERE BE-FORE?
UM... NO, IT'S MY FIRST TIME. UM... MY FIRST TIME HERE, I MEAN.
OKAY. WELL, HEAD DOWN THE HALLWAY AND FIND AN OPEN ROOM. YOU'LL FIND EVERY-THING YOU'LL NEED IN THERE. AND, WHEN YOU'RE DONE, YOU BRING YOUR DEPOSIT BACK TO ME.
WHAT ARE YOU SO WORRIED ABOUT? IT'S JUST MASTURBATION. EVERYBODY DOES IT.
YES, BUT EVERYBODY LIES ABOUT IT!
ONCE YOU'RE IN HERE, THERE'S NO MORE PRETENDING.
BESIDES WHICH, WHAT AM I MEANT TO DO WITH THIS LITTLE BOTTLE? I MEAN, JUST LOGISTICALLY.
AT WHAT POSSIBLE ANGLE...?

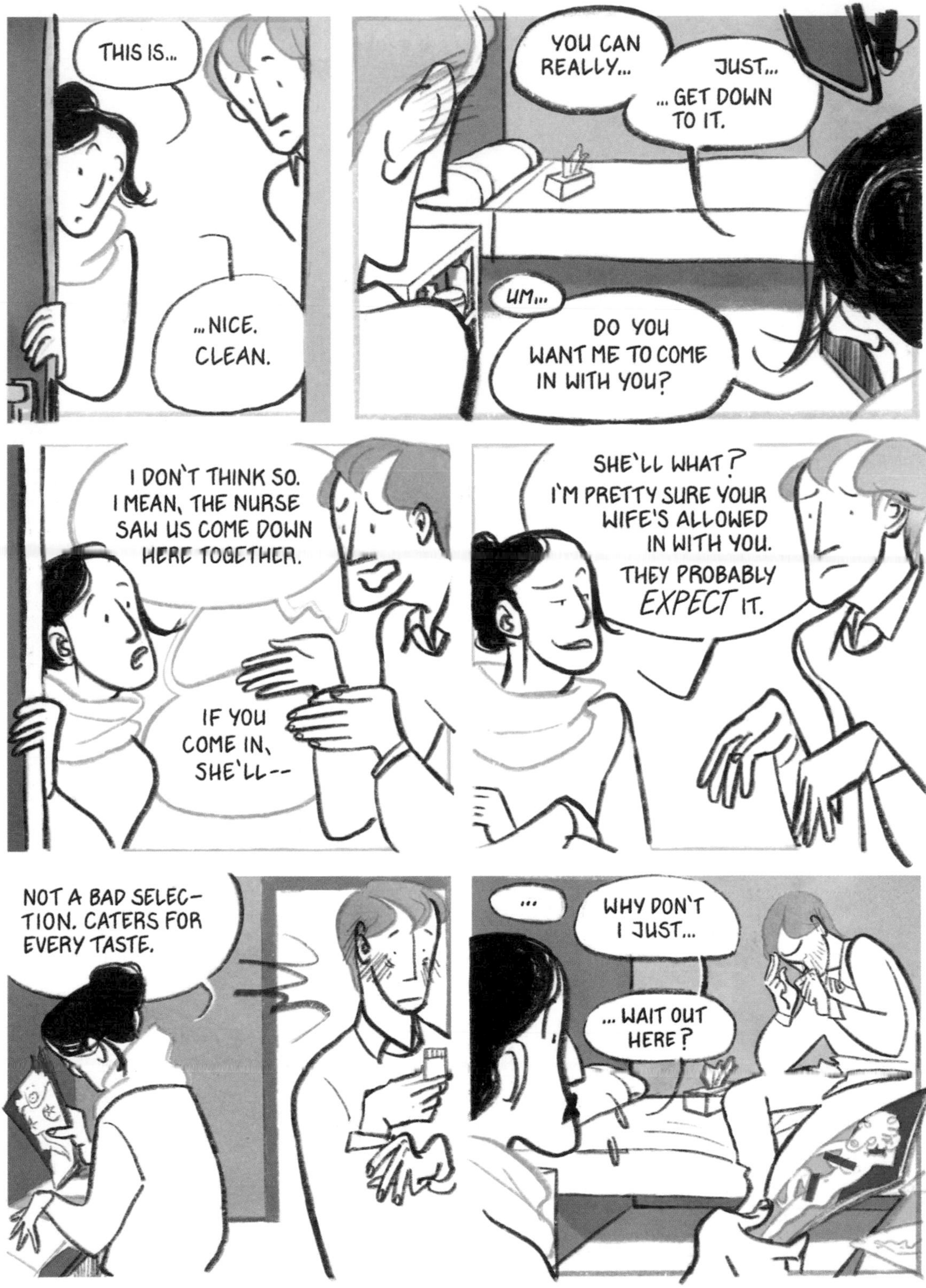
THIS IS...
...NICE. CLEAN.
YOU CAN REALLY...
JUST...
... GET DOWN TO IT.
UM...
DO YOU WANT ME TO COME IN WITH YOU?
I DON'T THINK SO. I MEAN, THE NURSE SAW US COME DOWN HERE TOGETHER.
IF YOU COME IN, SHE'LL--
SHE'LL WHAT? I'M PRETTY SURE YOUR WIFE'S ALLOWED IN WITH YOU. THEY PROBABLY EXPECT IT.
NOT A BAD SELEC-TION. CATERS FOR EVERY TASTE.
...
WHY DON'T I JUST...
... WAIT OUT HERE?

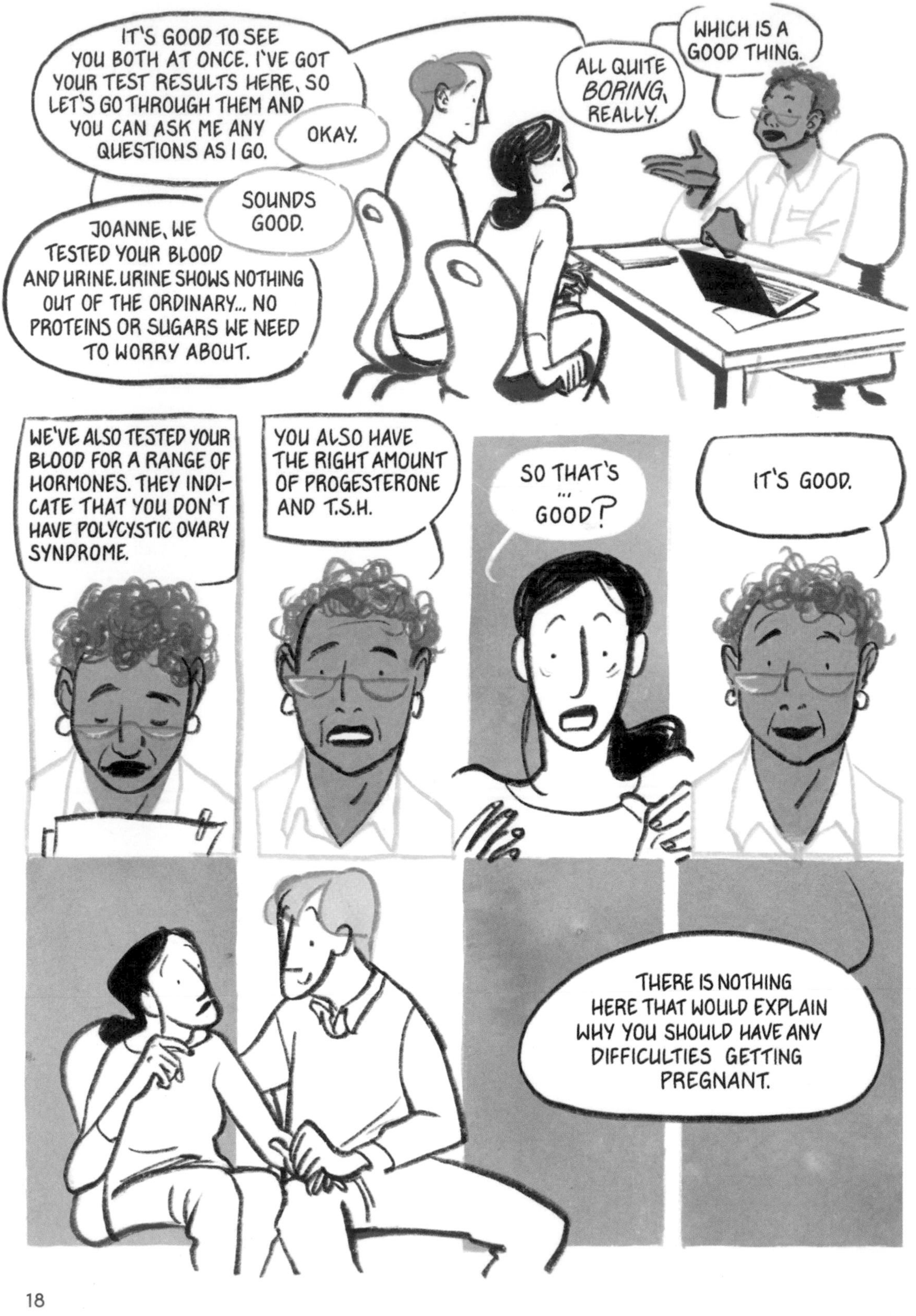
IT'S GOOD TO SEE YOU BOTH AT ONCE. I'VE GOT YOUR TEST RESULTS HERE, SO LET'S GO THROUGH THEM AND YOU CAN ASK ME ANY QUESTIONS AS I GO.
OKAY.
ALL QUITE *BORING*, REALLY.
WHICH IS A GOOD THING.
SOUNDS GOOD.
JOANNE, WE TESTED YOUR BLOOD AND URINE. URINE SHOWS NOTHING OUT OF THE ORDINARY... NO PROTEINS OR SUGARS WE NEED TO WORRY ABOUT.
WE'VE ALSO TESTED YOUR BLOOD FOR A RANGE OF HORMONES. THEY INDICATE THAT YOU DON'T HAVE POLYCYSTIC OVARY SYNDROME.
YOU ALSO HAVE THE RIGHT AMOUNT OF PROGESTERONE AND T.S.H.
SO THAT'S ... GOOD?
IT'S GOOD.
THERE IS NOTHING HERE THAT WOULD EXPLAIN WHY YOU SHOULD HAVE ANY DIFFICULTIES GETTING PREGNANT.

UNFORTUNATELY...

...CONRAD, YOUR TESTS WERE LESS PROMISING.

WHEN WE LOOK AT SPERM, WE CONSIDER A COUPLE OF THINGS.

ONE IS THE QUALITY.

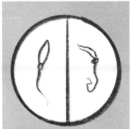

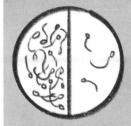

WE CHECK THE SIZE AND SHAPE OF THE SPERM THEMSELVES

– WHAT WE CALL THE MORPHOLOGY–

AND THE MOVEMENT AND ACTIVITY OF THE CELLS, WHICH WE CALL THE MOTILITY. BOTH OF THESE ARE FINE.

ANOTHER THING WE LOOK AT, THOUGH, IS THE QUANTITY:

YOUR SPERM COUNT.

WE'D BE HOPING TO SEE BETWEEN 20 AND 300 MILLION PER MILLILITRE.

YOU'RE CLOSER TO 10 MILLION.

I'M NOT SAYING THIS IS DEFINITIVE.

I MEAN, A LOWER-THAN-IDEAL SPERM COUNT *MIGHT* OR *MIGHT NOT* ACCOUNT FOR THE DIFFICULTIES YOU'VE BEEN EXPERIENCING--

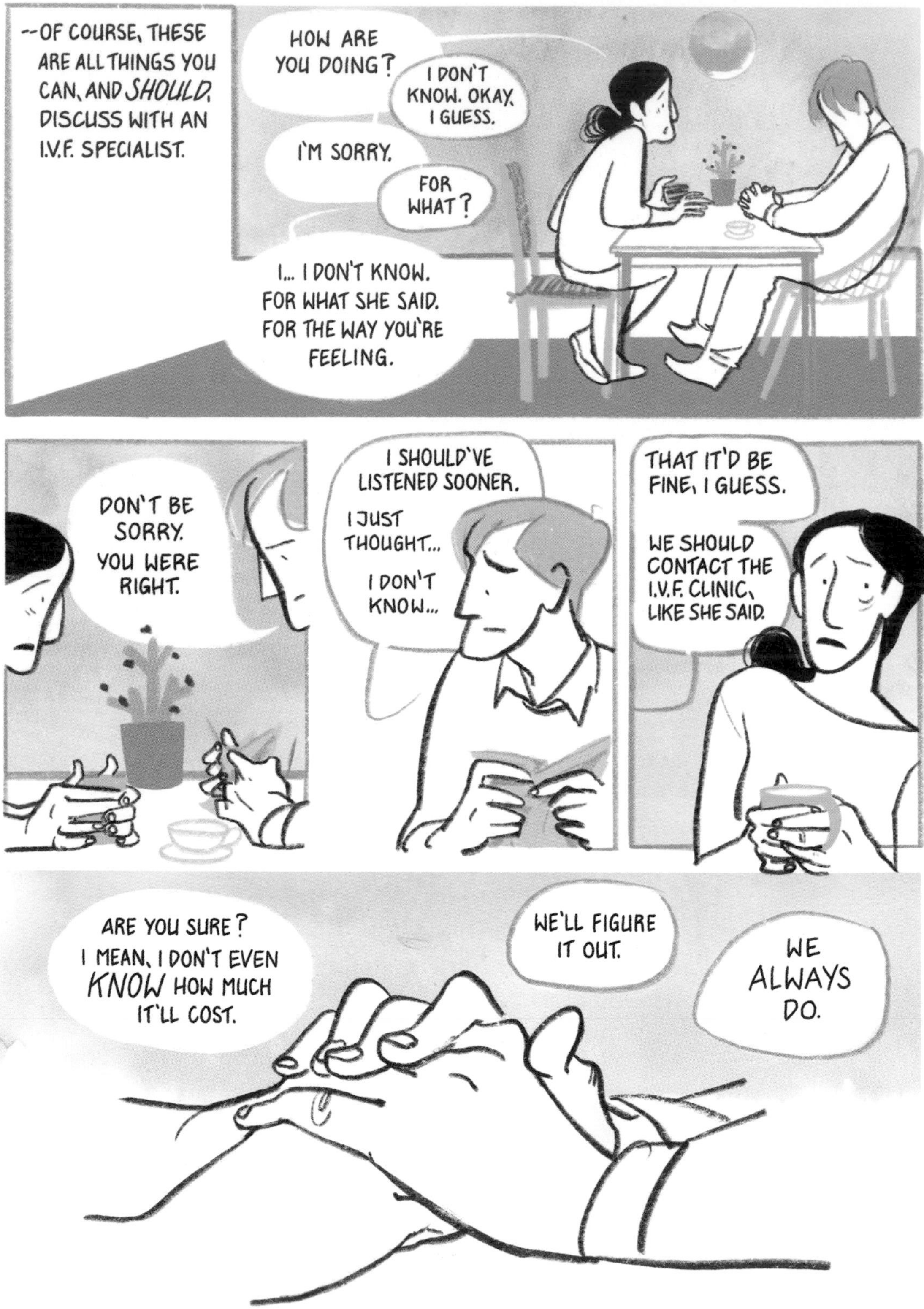
--OF COURSE, THESE ARE ALL THINGS YOU CAN, AND *SHOULD*, DISCUSS WITH AN I.V.F. SPECIALIST.
HOW ARE YOU DOING?
I DON'T KNOW. OKAY. I GUESS.
I'M SORRY.
FOR WHAT?
I... I DON'T KNOW. FOR WHAT SHE SAID. FOR THE WAY YOU'RE FEELING.
DON'T BE SORRY. YOU WERE RIGHT.
I SHOULD'VE LISTENED SOONER.
I JUST THOUGHT...
I DON'T KNOW...
THAT IT'D BE FINE, I GUESS.
WE SHOULD CONTACT THE I.V.F. CLINIC, LIKE SHE SAID.
ARE YOU SURE? I MEAN, I DON'T EVEN *KNOW* HOW MUCH IT'LL COST.
WE'LL FIGURE IT OUT.
WE ALWAYS DO.

CHAPTER 2:

THE PROCESS BEGINS

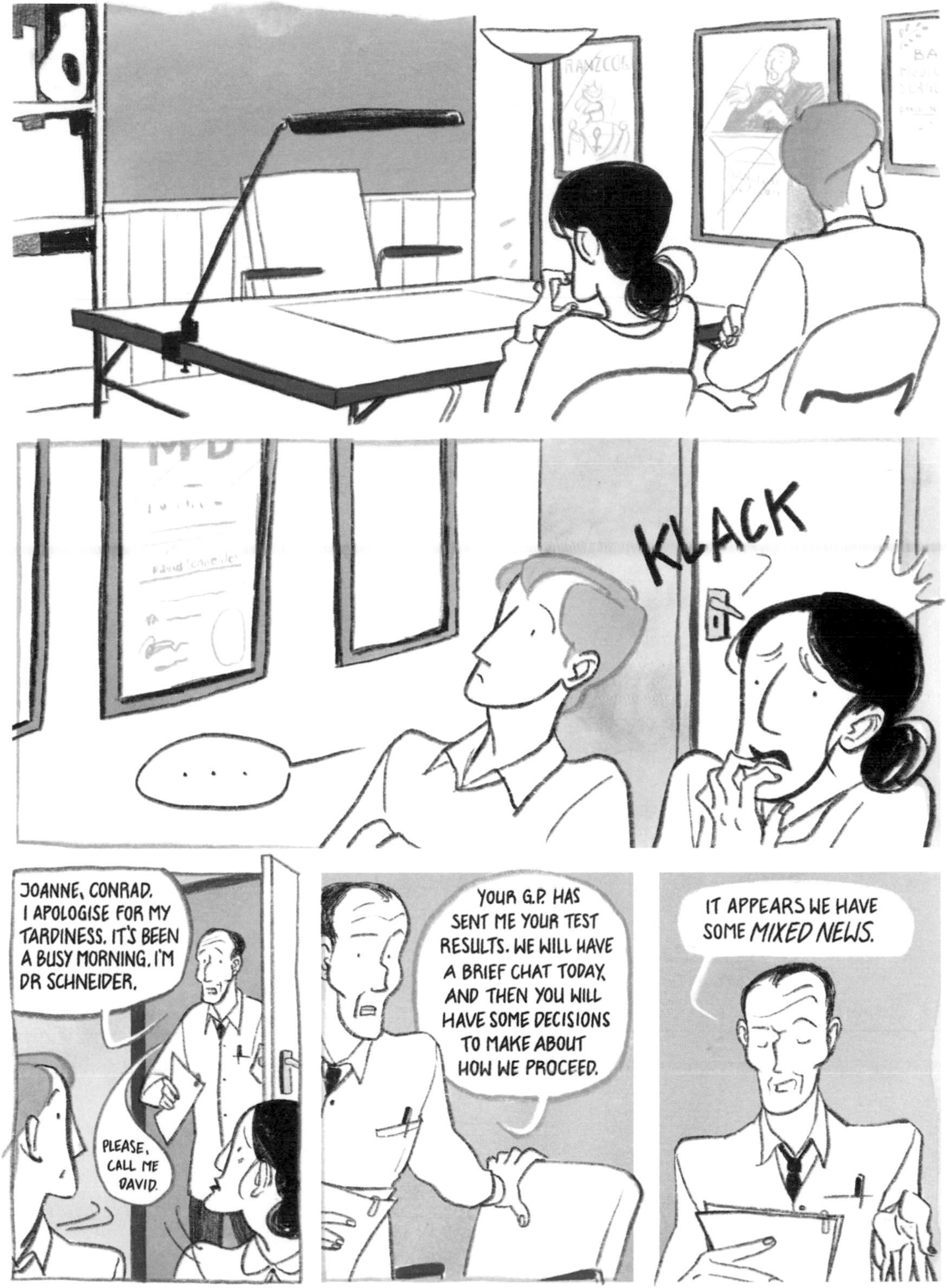
KLACK
. . .
JOANNE, CONRAD. I APOLOGISE FOR MY TARDINESS. IT'S BEEN A BUSY MORNING. I'M DR SCHNEIDER.
PLEASE, CALL ME DAVID.
YOUR G.P. HAS SENT ME YOUR TEST RESULTS. WE WILL HAVE A BRIEF CHAT TODAY, AND THEN YOU WILL HAVE SOME DECISIONS TO MAKE ABOUT HOW WE PROCEED.
IT APPEARS WE HAVE SOME *MIXED NEWS*.

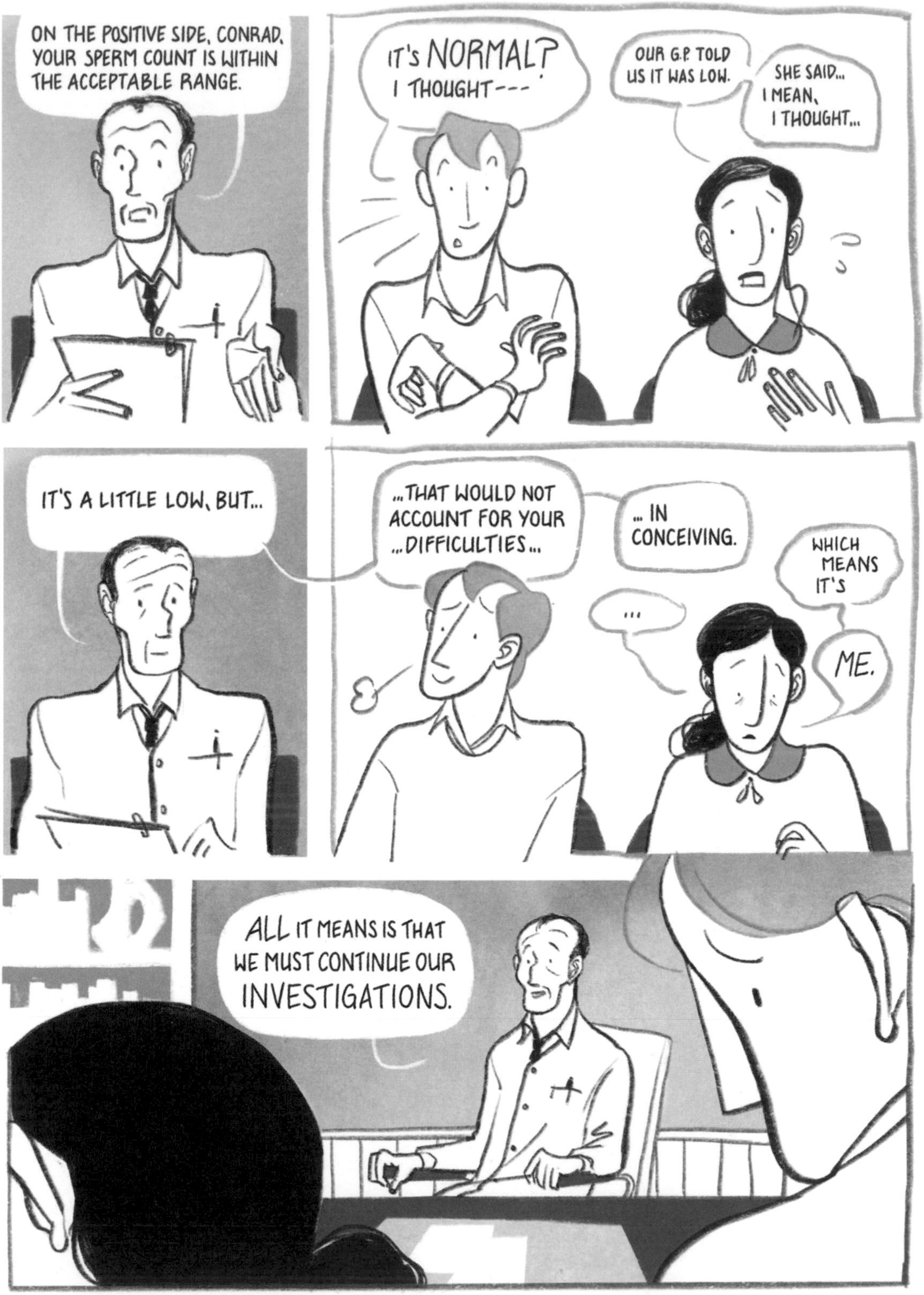
ON THE POSITIVE SIDE, CONRAD, YOUR SPERM COUNT IS WITHIN THE ACCEPTABLE RANGE.
IT'S NORMAL? I THOUGHT---
OUR G.P. TOLD US IT WAS LOW.
SHE SAID... I MEAN, I THOUGHT...
IT'S A LITTLE LOW, BUT...
...THAT WOULD NOT ACCOUNT FOR YOUR ...DIFFICULTIES...
... IN CONCEIVING.
...
WHICH MEANS IT'S
ME.
ALL IT MEANS IS THAT WE MUST CONTINUE OUR INVESTIGATIONS.

JOANNE, I'M NOT SURE HOW MUCH DR SCHNEIDER TOLD YOU ABOUT WHY WE PERFORM THIS PROCEDURE---

HE DIDN'T, REALLY. HE JUST SAID HE NEEDED TO GET SOME MORE ANSWERS...

WELL, THIS IS CALLED A HYSTEROSALPINGO-GRAM, OR H.S.G. FOR SHORT.

NOW THAT WE'VE INSERTED THE SPECULUM, I'M GOING TO FEED A CANULA THROUGH YOUR CERVIX.

SOUNDS LIKE FUN.

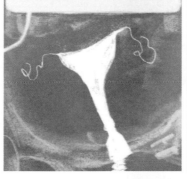

NOW, I'M INJECTING CONTRAST DYE INTO YOUR UTERINE CAVITY – YOU CAN SEE IT THERE: THE WHITE SHAPE IN THE CENTRE.

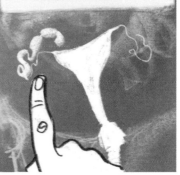

AS I CONTINUE TO PUSH THE DYE THROUGH YOUR CERVIX, WE SHOULD SEE YOUR FALLOPIAN TUBES LIGHT UP.

NOW, WE SHOULD START TO SEE THE TUBES SPILLING THE DYE INTO THE ABDOMINAL CAVITY.

OUCH!

ISN'T IT MEANT TO?

THIS HURTS?

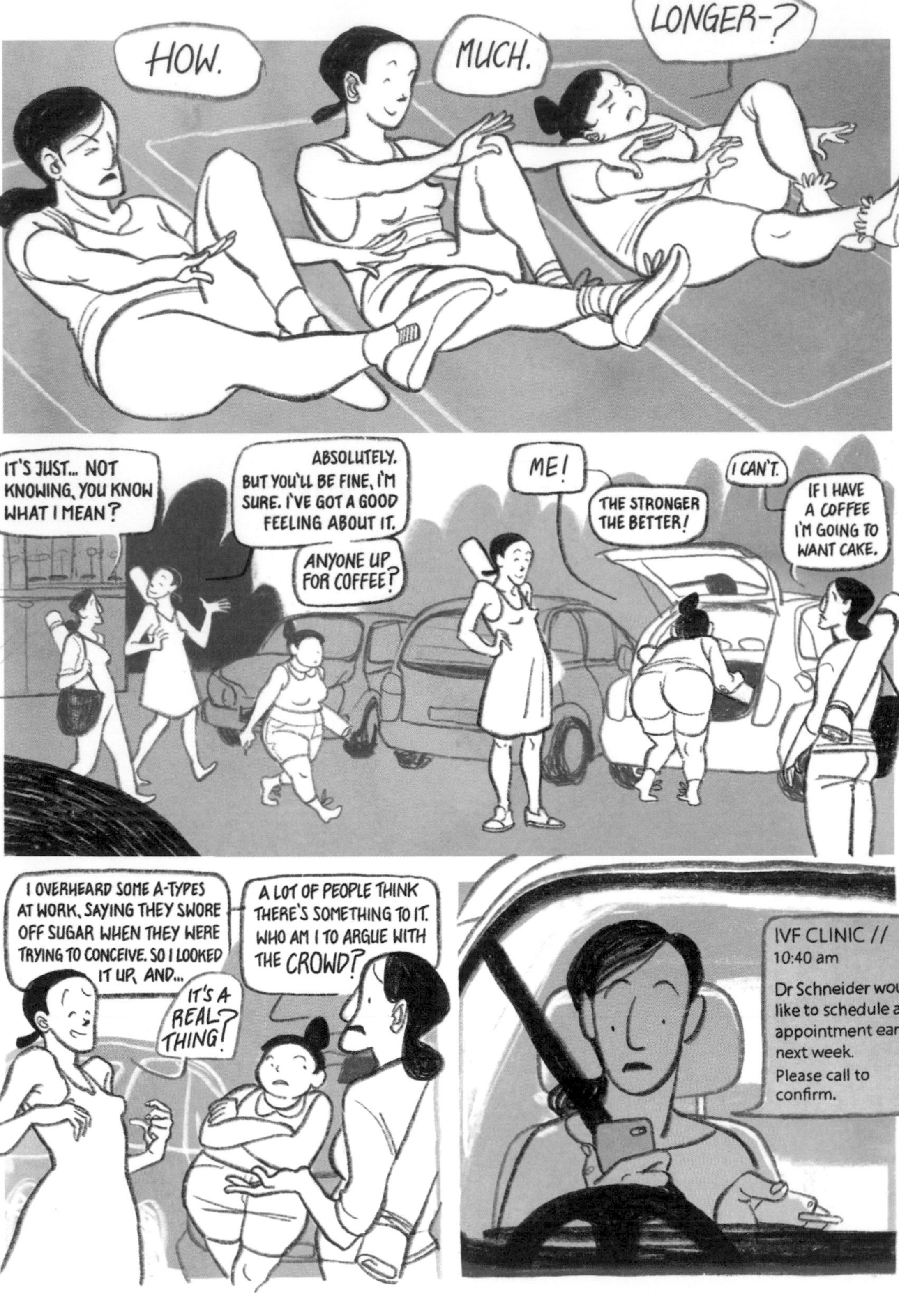
HOW.
MUCH.
LONGER-?
IT'S JUST... NOT KNOWING, YOU KNOW WHAT I MEAN?
ABSOLUTELY. BUT YOU'LL BE FINE, I'M SURE. I'VE GOT A GOOD FEELING ABOUT IT.
ANYONE UP FOR COFFEE?
ME!
THE STRONGER THE BETTER!
I CAN'T.
IF I HAVE A COFFEE I'M GOING TO WANT CAKE.
I OVERHEARD SOME A-TYPES AT WORK, SAYING THEY SWORE OFF SUGAR WHEN THEY WERE TRYING TO CONCEIVE. SO I LOOKED IT UP, AND...
A LOT OF PEOPLE THINK THERE'S SOMETHING TO IT. WHO AM I TO ARGUE WITH THE CROWD?
IT'S A REAL THING?
IVF CLINIC //
10:40 am
Dr Schneider would like to schedule an appointment early next week.
Please call to confirm.

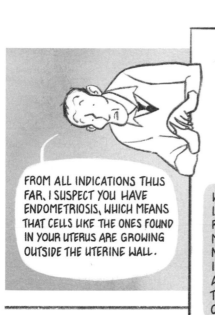

FROM ALL INDICATIONS THUS FAR, I SUSPECT YOU HAVE ENDOMETRIOSIS, WHICH MEANS THAT CELLS LIKE THE ONES FOUND IN YOUR UTERUS ARE GROWING OUTSIDE THE UTERINE WALL.

FOR AROUND ONE-THIRD OF WOMEN WHO SUFFER FROM THE CONDITION, IT MAKES IT HARDER FOR THEM TO CONCEIVE. THAT MAY BE WHAT IS HAPPENING TO YOU.

WHAT I WOULD LIKE TO DO IS BOOK YOU FOR A LAPA-ROSCOPY. THIS WILL ALLOW ME TO CONFIRM MY DIAG-NOSIS, AND — ASSUMING I AM RIGHT — TO CLEAN AWAY ANY ENDOMETRIAL TISSUE I FIND. IS THAT CLEAR?

WHEN YOU SAY 'CLEAN AWAY', YOU MEAN *CUT*.

YES.

YOU *HFARD* SCHNEIDER: IT'S NOT A DANGEROUS OPER-ATION. YOU'LL BE *OKAY*, JO.

...

... I'VE BEEN SO *GOOD* RECENTLY: CHUGGING DOWN VITAMINS AND WATER, HITTING THE GYM... SAYING 'NO' TO ANYTHING I ACTUALLY *WANT* TO EAT!

ALL ON THE OFF-CHANCE THAT IT WOULD MAKE A DIFFERENCE.

WELL, AT LEAST THERE'S AN UPSIDE, THEN.

YOU'RE COMING WITH ME.

FEEL BETTER?

I'M GETTING THERE.

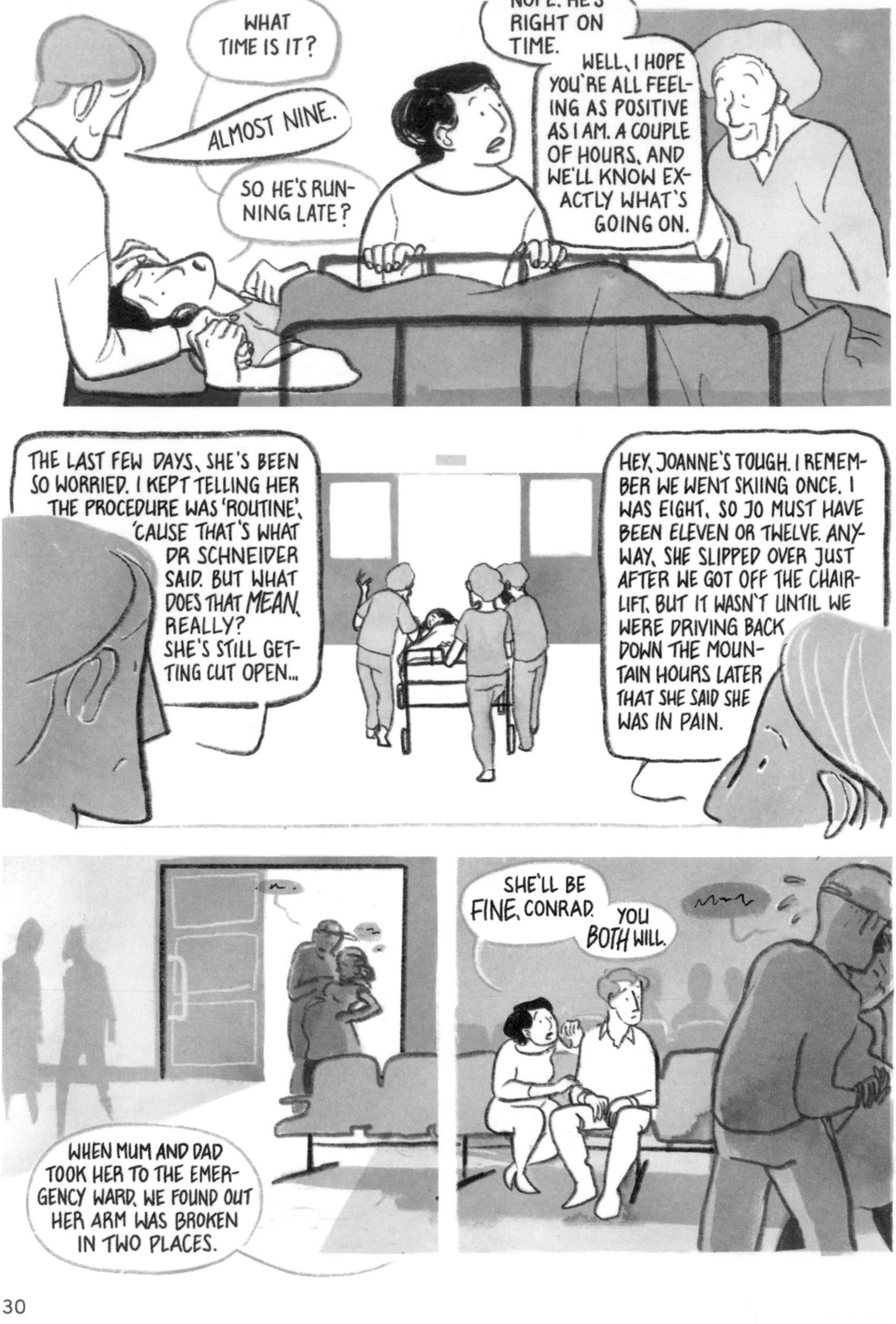
WHAT TIME IS IT?
ALMOST NINE.
SO HE'S RUNNING LATE?
NOPE. HE'S RIGHT ON TIME.
WELL, I HOPE YOU'RE ALL FEELING AS POSITIVE AS I AM. A COUPLE OF HOURS, AND WE'LL KNOW EXACTLY WHAT'S GOING ON.
THE LAST FEW DAYS, SHE'S BEEN SO WORRIED. I KEPT TELLING HER THE PROCEDURE WAS 'ROUTINE', 'CAUSE THAT'S WHAT DR SCHNEIDER SAID. BUT WHAT DOES THAT MEAN, REALLY? SHE'S STILL GETTING CUT OPEN...
HEY, JOANNE'S TOUGH. I REMEMBER WE WENT SKIING ONCE. I WAS EIGHT, SO JO MUST HAVE BEEN ELEVEN OR TWELVE. ANYWAY, SHE SLIPPED OVER JUST AFTER WE GOT OFF THE CHAIRLIFT, BUT IT WASN'T UNTIL WE WERE DRIVING BACK DOWN THE MOUNTAIN HOURS LATER THAT SHE SAID SHE WAS IN PAIN.
WHEN MUM AND DAD TOOK HER TO THE EMERGENCY WARD, WE FOUND OUT HER ARM WAS BROKEN IN TWO PLACES.
SHE'LL BE FINE, CONRAD.
YOU BOTH WILL.

FOUR HOURS LATER...

JO, ARE YOU AWAKE?

ER M...

I'M NOT... I WAS...

I THINK SO.

THE LIGHT! WHY IS IT SO BRIGHT?

IT JUST FEELS THAT WAY, JO. YOU'VE BEEN UNDER ANAESTHETIC, REMEMBER? TRY TO RELAX.

DR SCHNEIDER SAID HE'D BE HERE IN A SEC.

MAYBE HAVING KIDS ISN'T EVERYTHING. WHAT IF WE HAVE A KID, AND WE DON'T LIKE IT? WE'LL BE STUCK WITH IT!

FOR YEARS!

I KNOW YOU DON'T MEAN THAT.

HELLO?

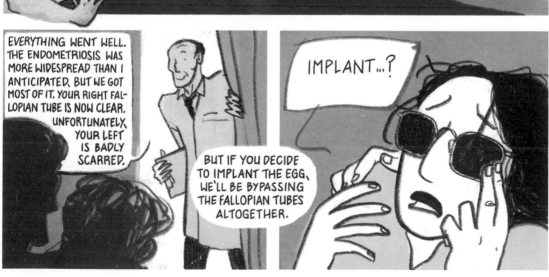

EVERYTHING WENT WELL. THE ENDOMETRIOSIS WAS MORE WIDESPREAD THAN I ANTICIPATED, BUT WE GOT MOST OF IT. YOUR RIGHT FALLOPIAN TUBE IS NOW CLEAR. UNFORTUNATELY, YOUR LEFT IS BADLY SCARRED.

BUT IF YOU DECIDE TO IMPLANT THE EGG, WE'LL BE BYPASSING THE FALLOPIAN TUBES ALTOGETHER.

IMPLANT...?

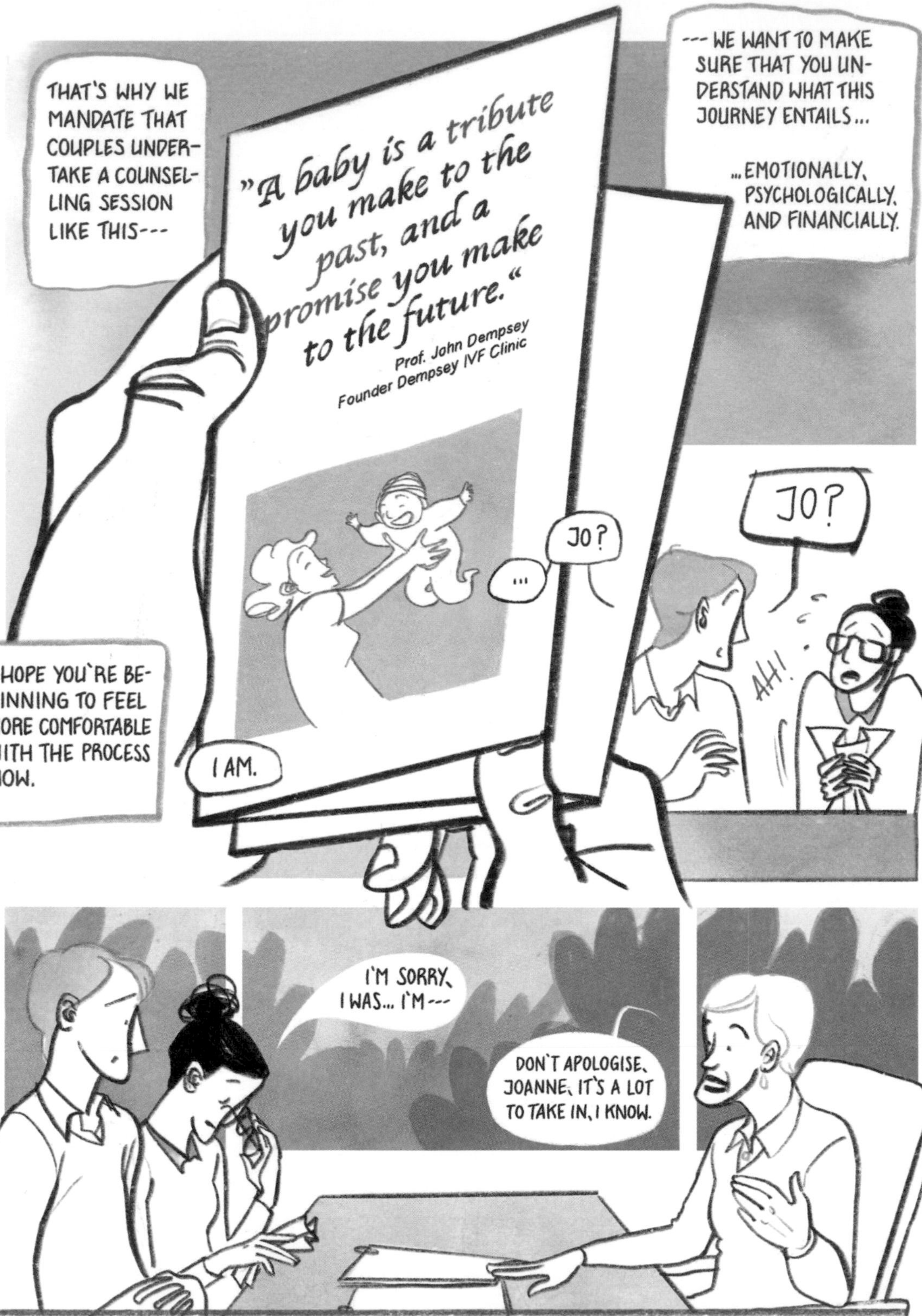
THAT'S WHY WE MANDATE THAT COUPLES UNDERTAKE A COUNSELLING SESSION LIKE THIS---
"A baby is a tribute you make to the past, and a promise you make to the future."
Prof. John Dempsey
Founder Dempsey IVF Clinic
--- WE WANT TO MAKE SURE THAT YOU UNDERSTAND WHAT THIS JOURNEY ENTAILS...
...EMOTIONALLY, PSYCHOLOGICALLY, AND FINANCIALLY.
I HOPE YOU'RE BEGINNING TO FEEL MORE COMFORTABLE WITH THE PROCESS NOW.
I AM.
...
JO?
JO?
AH!
I'M SORRY, I WAS... I'M ---
DON'T APOLOGISE, JOANNE, IT'S A LOT TO TAKE IN, I KNOW.

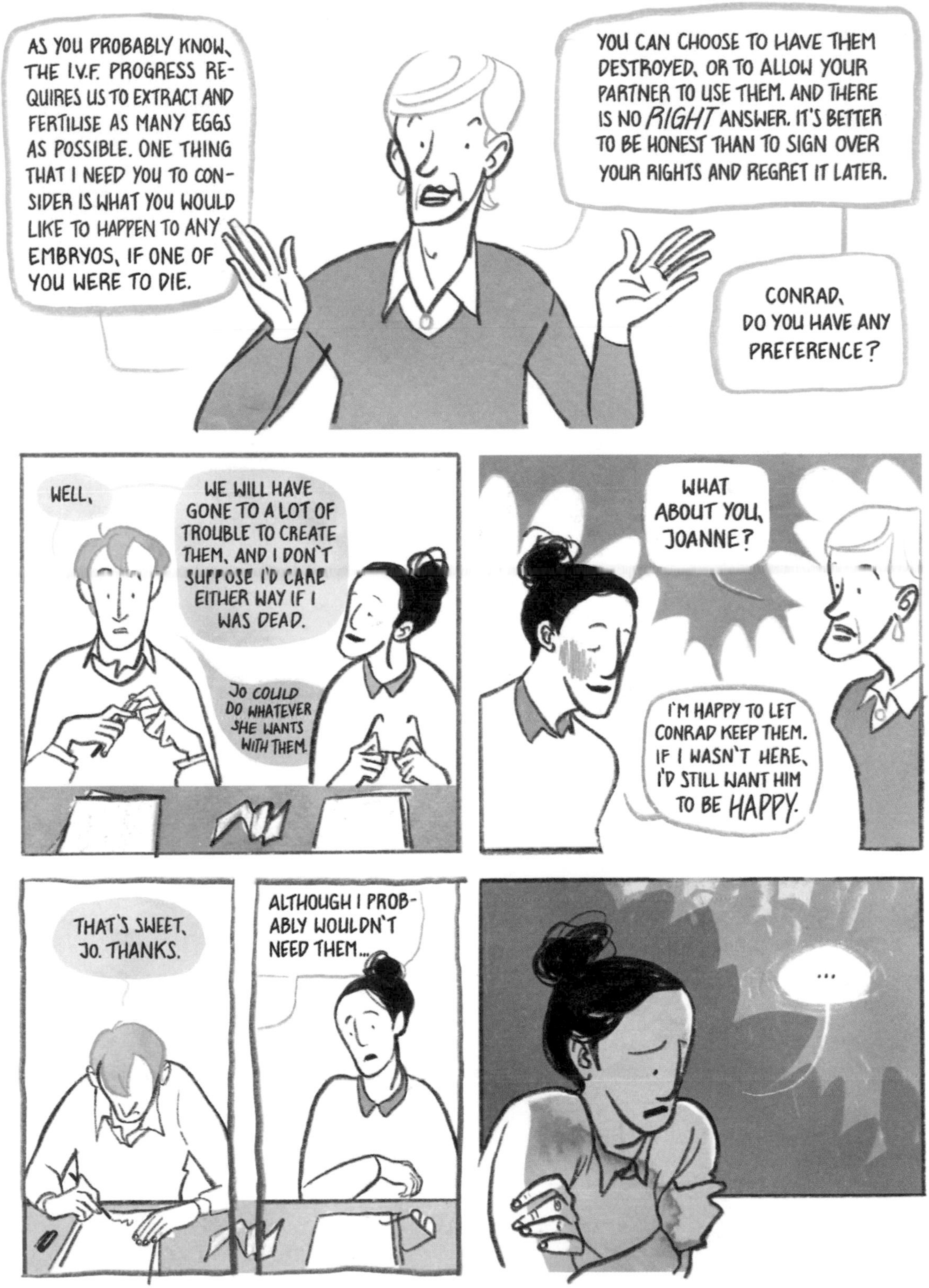
AS YOU PROBABLY KNOW, THE I.V.F. PROGRESS REQUIRES US TO EXTRACT AND FERTILISE AS MANY EGGS AS POSSIBLE. ONE THING THAT I NEED YOU TO CONSIDER IS WHAT YOU WOULD LIKE TO HAPPEN TO ANY EMBRYOS, IF ONE OF YOU WERE TO DIE.
YOU CAN CHOOSE TO HAVE THEM DESTROYED, OR TO ALLOW YOUR PARTNER TO USE THEM. AND THERE IS NO *RIGHT* ANSWER. IT'S BETTER TO BE HONEST THAN TO SIGN OVER YOUR RIGHTS AND REGRET IT LATER.
CONRAD, DO YOU HAVE ANY PREFERENCE?
WELL,
WE WILL HAVE GONE TO A LOT OF TROUBLE TO CREATE THEM, AND I DON'T SUPPOSE I'D CARE EITHER WAY IF I WAS DEAD.
JO COULD DO WHATEVER SHE WANTS WITH THEM.
WHAT ABOUT YOU, JOANNE?
I'M HAPPY TO LET CONRAD KEEP THEM. IF I WASN'T HERE, I'D STILL WANT HIM TO BE HAPPY.
THAT'S SWEET, JO. THANKS.
ALTHOUGH I PROBABLY WOULDN'T NEED THEM...
...

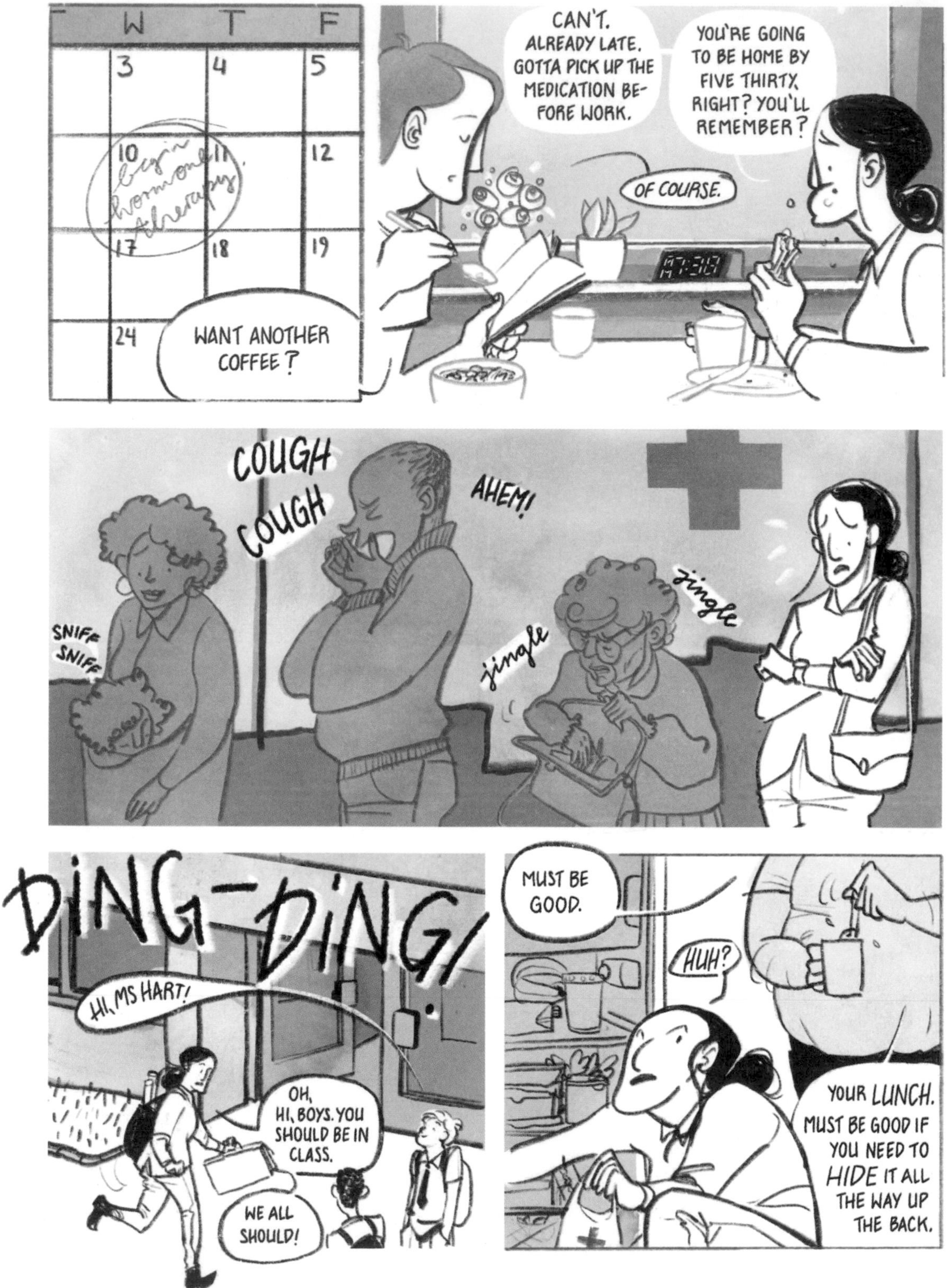

W
T
F
3
4
5
10
11
12
begin hormone therapy
17
18
19
24
WANT ANOTHER COFFEE?
CAN'T. ALREADY LATE. GOTTA PICK UP THE MEDICATION BE-FORE WORK.
YOU'RE GOING TO BE HOME BY FIVE THIRTY, RIGHT? YOU'LL REMEMBER?
OF COURSE.
COUGH
COUGH
AHEM!
SNIFF SNIFF
jingle
jingle
DING-DING!
HI, MS HART!
OH, HI, BOYS. YOU SHOULD BE IN CLASS.
WE ALL SHOULD!
MUST BE GOOD.
HUH?
YOUR LUNCH. MUST BE GOOD IF YOU NEED TO HIDE IT ALL THE WAY UP THE BACK.

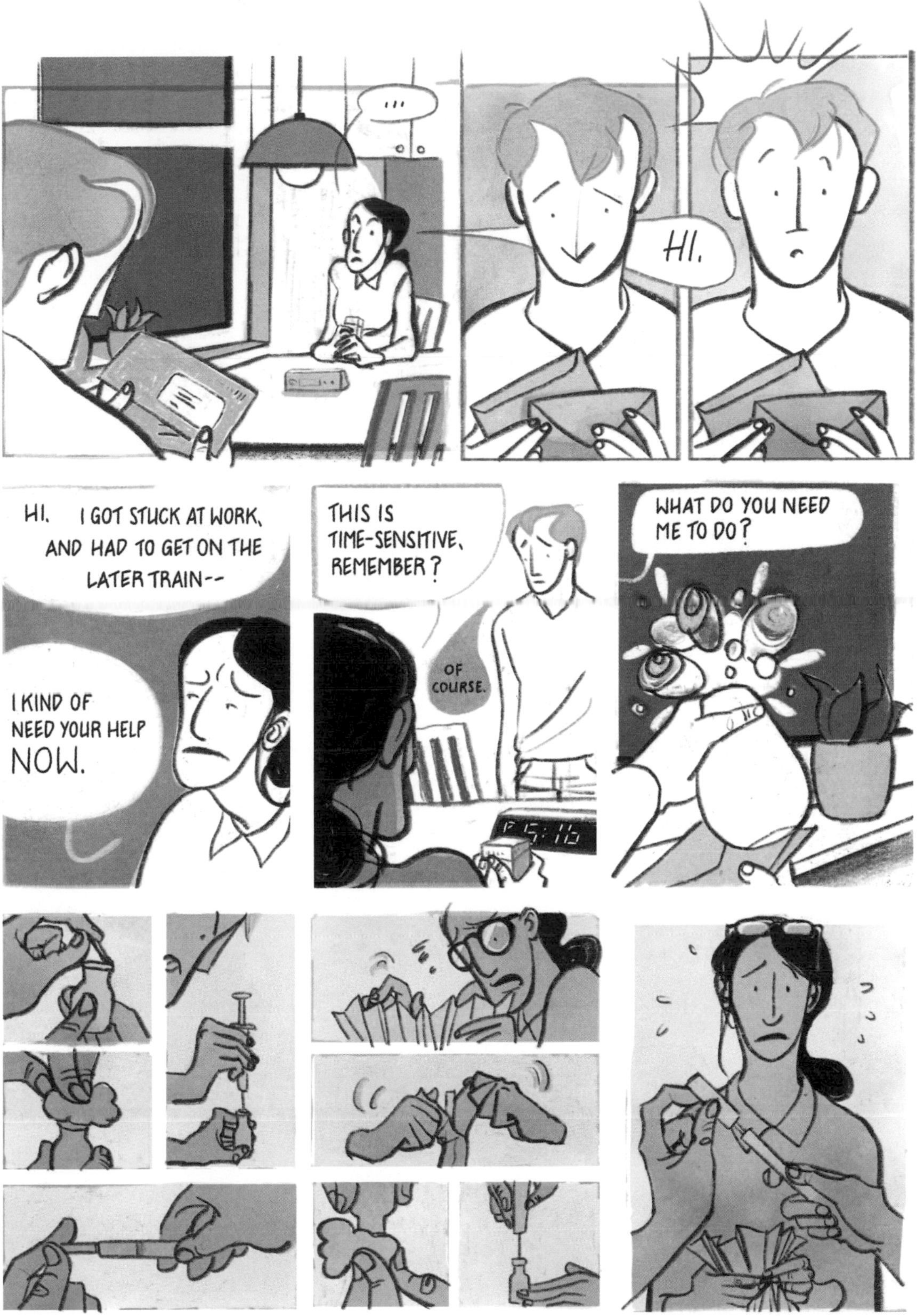
...
HI.
HI. I GOT STUCK AT WORK, AND HAD TO GET ON THE LATER TRAIN--
I KIND OF NEED YOUR HELP NOW.
THIS IS TIME-SENSITIVE, REMEMBER?
OF COURSE.
WHAT DO YOU NEED ME TO DO?

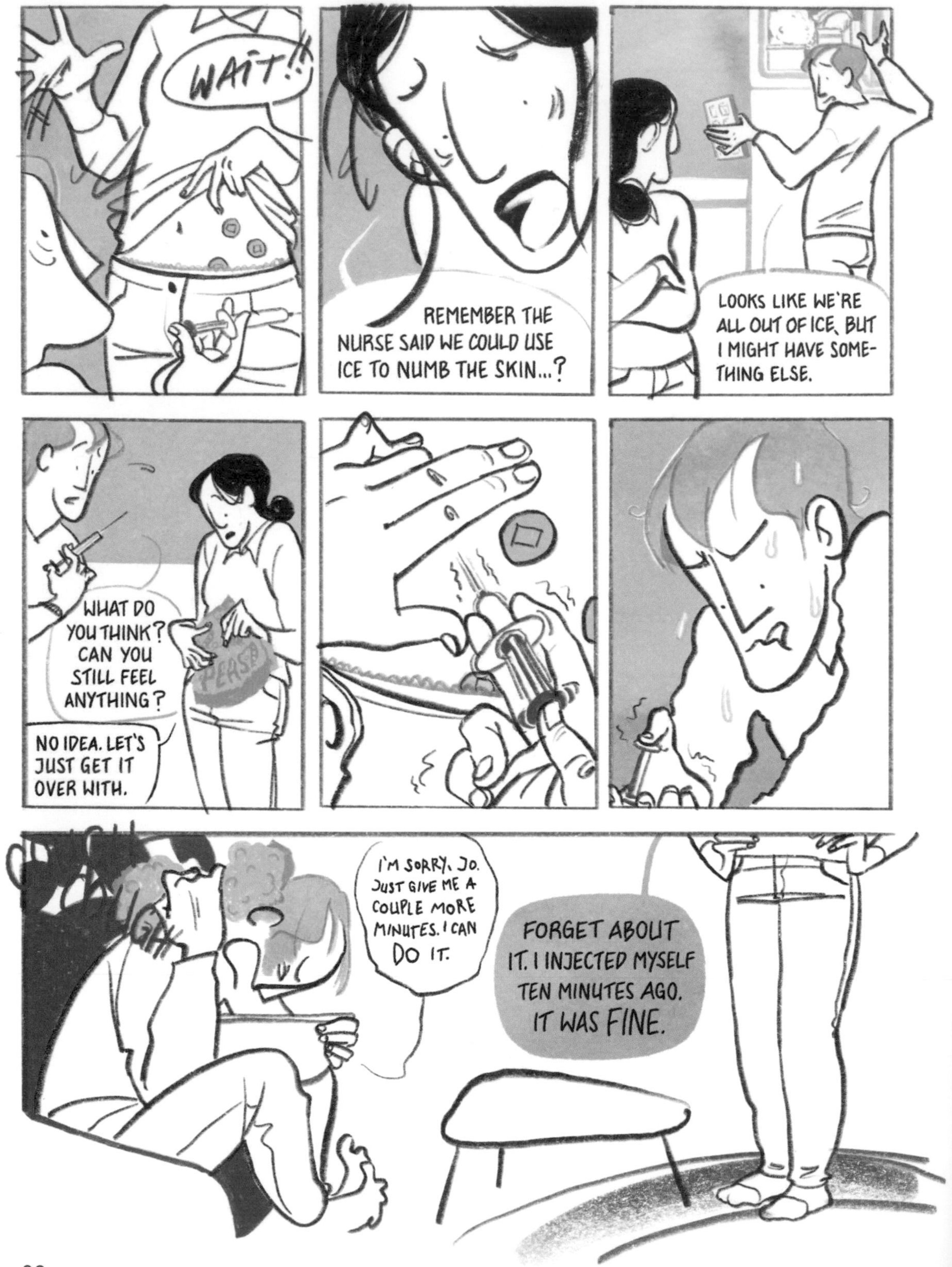
WAIT!!
REMEMBER THE NURSE SAID WE COULD USE ICE TO NUMB THE SKIN...?
LOOKS LIKE WE'RE ALL OUT OF ICE, BUT I MIGHT HAVE SOMETHING ELSE.
PEAS
WHAT DO YOU THINK? CAN YOU STILL FEEL ANYTHING?
NO IDEA. LET'S JUST GET IT OVER WITH.
I'M SORRY, JO. JUST GIVE ME A COUPLE MORE MINUTES. I CAN DO IT.
FORGET ABOUT IT. I INJECTED MYSELF TEN MINUTES AGO. IT WAS FINE.

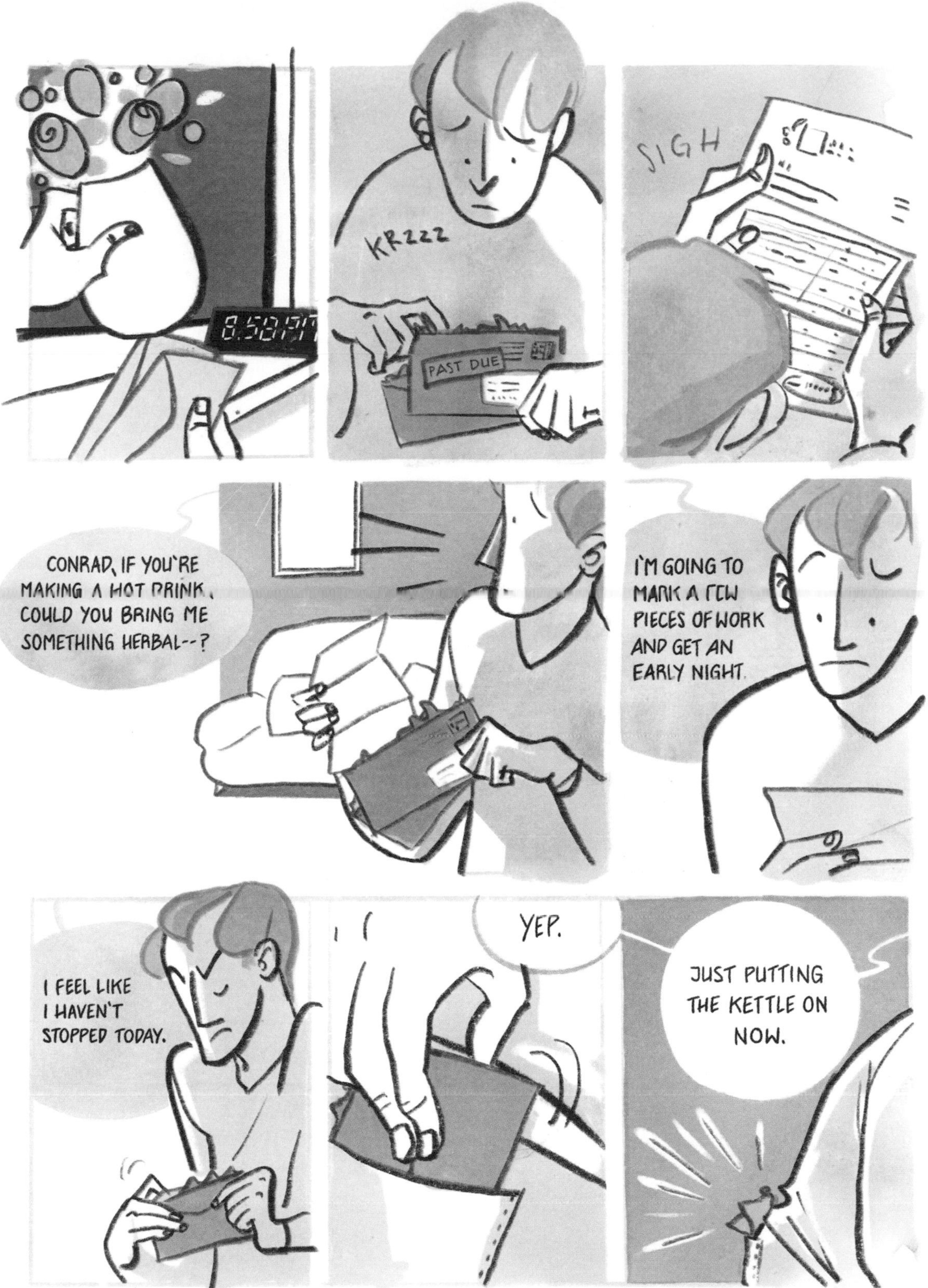
8:58PM
KRZZZ
PAST DUE
SIGH
CONRAD, IF YOU'RE MAKING A HOT DRINK, COULD YOU BRING ME SOMETHING HERBAL--?
I'M GOING TO MARK A FEW PIECES OF WORK AND GET AN EARLY NIGHT.
I FEEL LIKE I HAVEN'T STOPPED TODAY.
YEP.
JUST PUTTING THE KETTLE ON NOW.

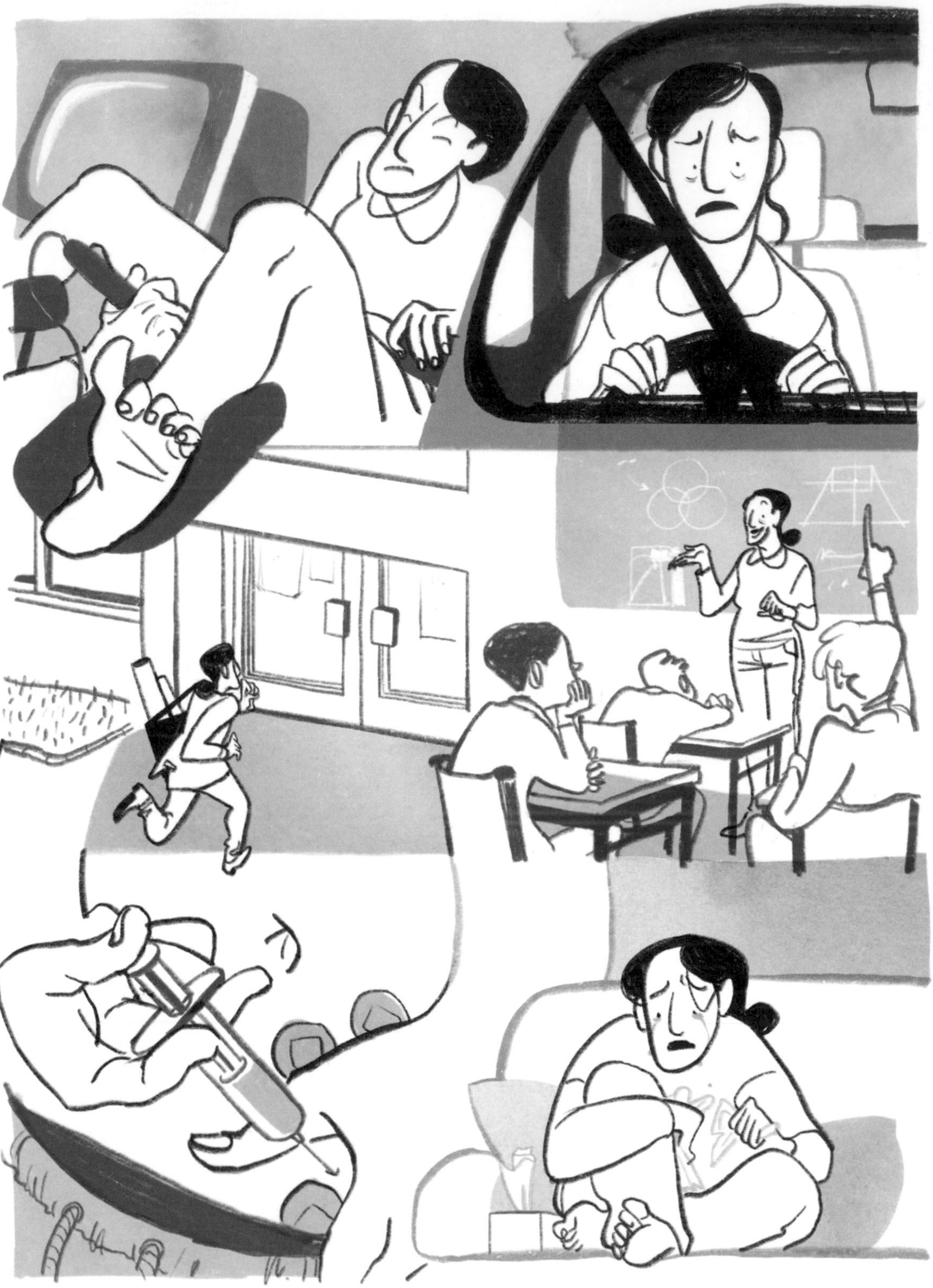

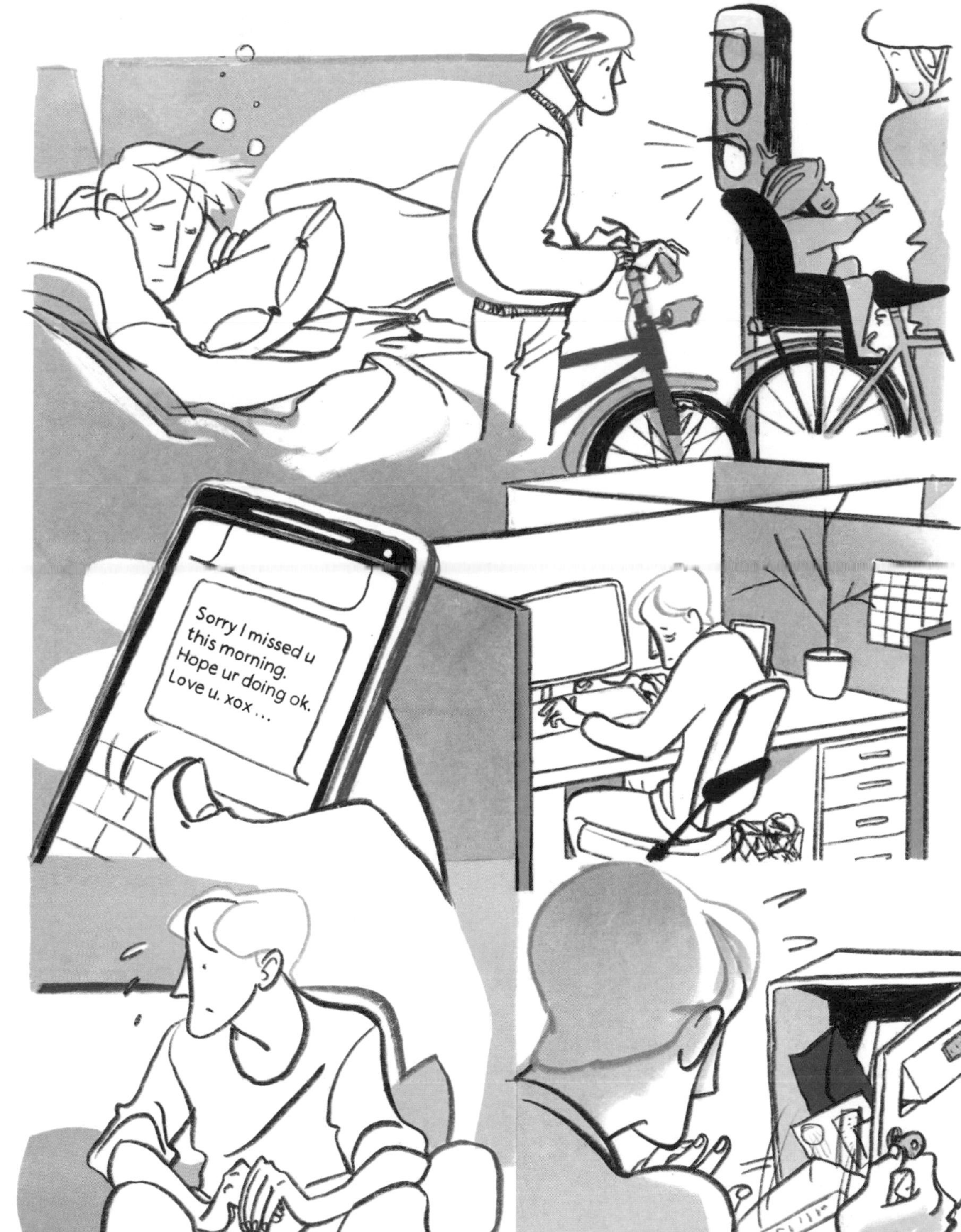
Sorry I missed u this morning. Hope ur doing ok. Love u. xox . . .

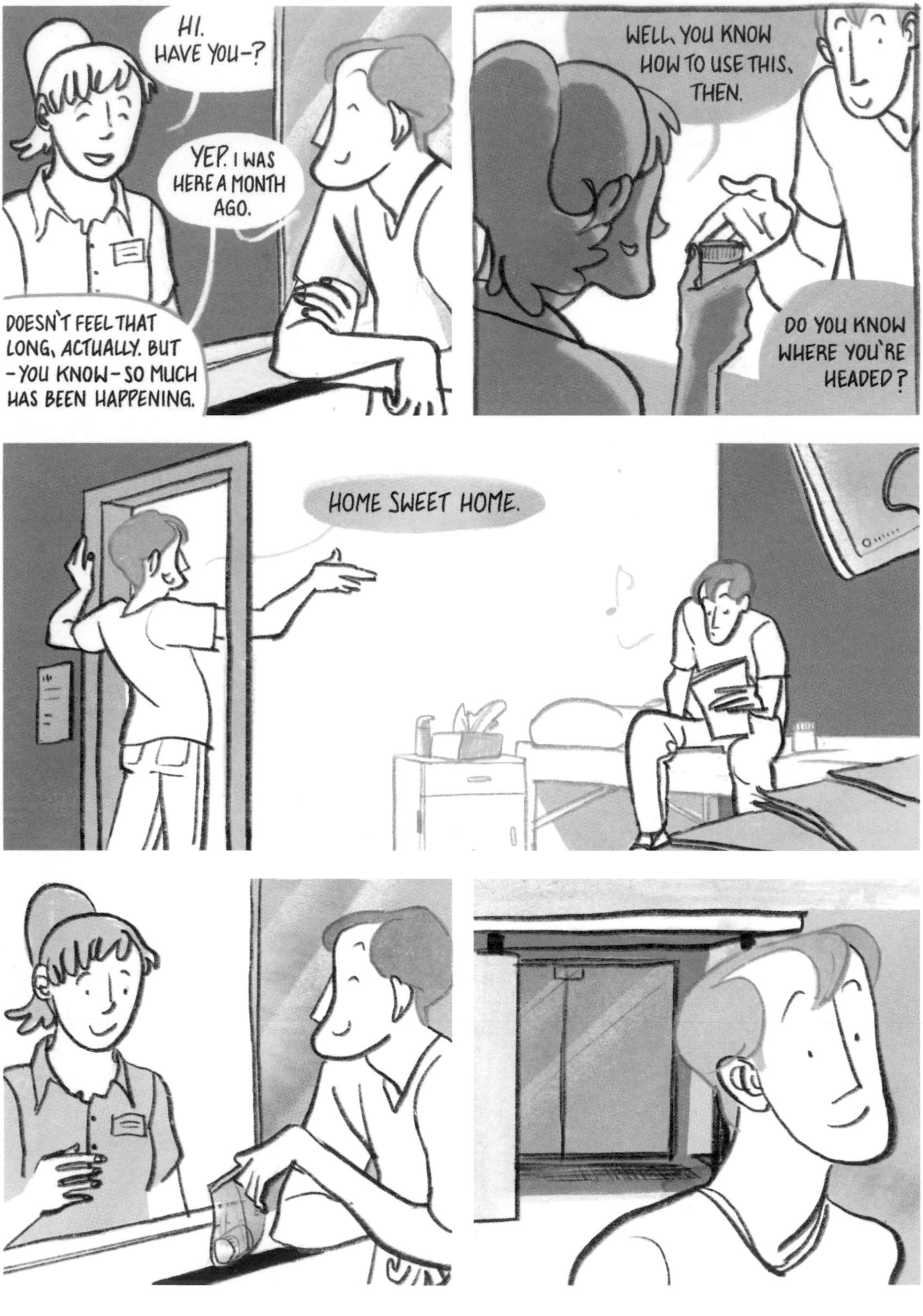
HI. HAVE YOU-?
YEP. I WAS HERE A MONTH AGO.
DOESN'T FEEL THAT LONG, ACTUALLY. BUT -YOU KNOW- SO MUCH HAS BEEN HAPPENING.
WELL, YOU KNOW HOW TO USE THIS, THEN.
DO YOU KNOW WHERE YOU'RE HEADED?
HOME SWEET HOME.

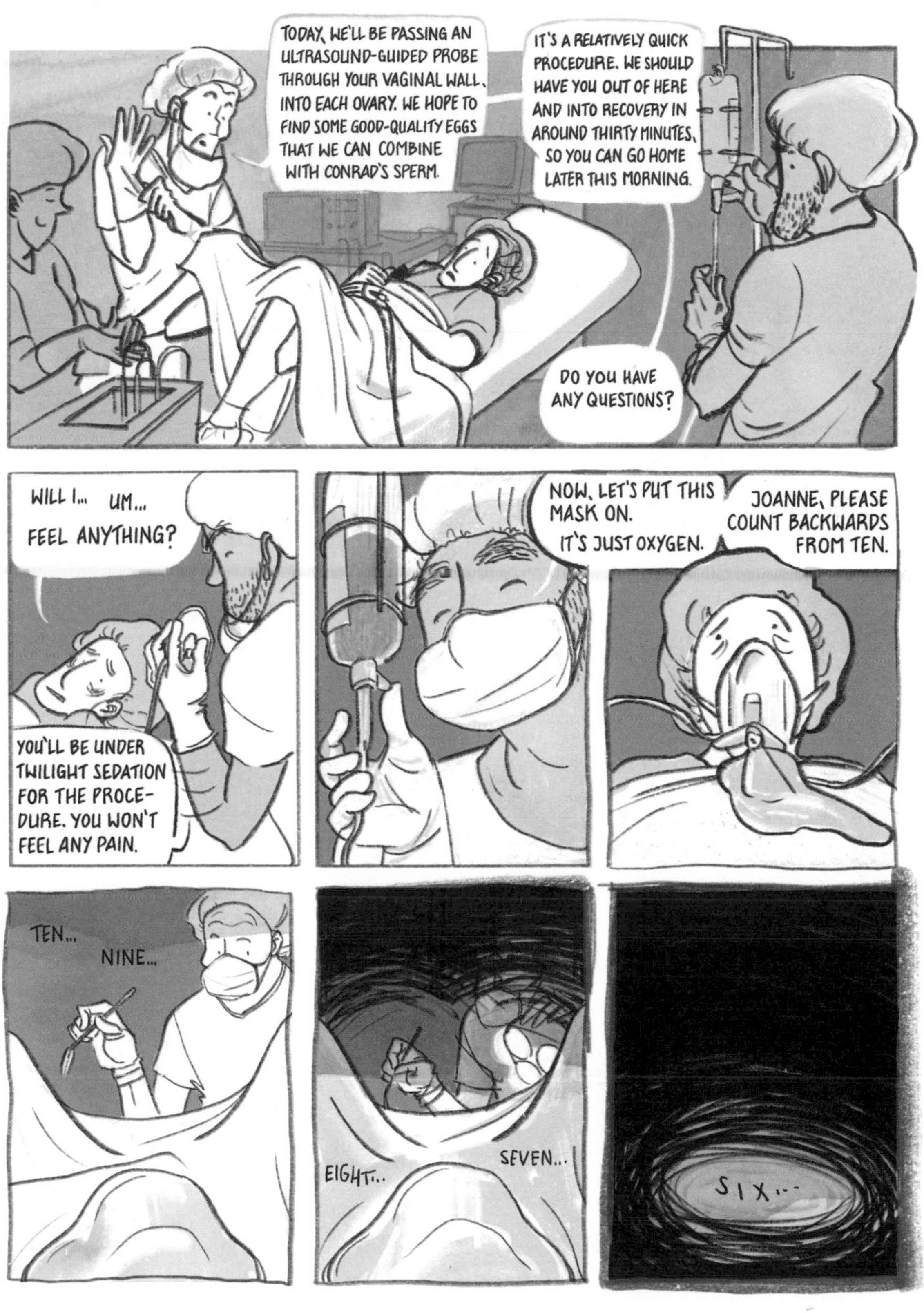

TODAY, WE'LL BE PASSING AN ULTRASOUND-GUIDED PROBE THROUGH YOUR VAGINAL WALL, INTO EACH OVARY. WE HOPE TO FIND SOME GOOD-QUALITY EGGS THAT WE CAN COMBINE WITH CONRAD'S SPERM.
IT'S A RELATIVELY QUICK PROCEDURE. WE SHOULD HAVE YOU OUT OF HERE AND INTO RECOVERY IN AROUND THIRTY MINUTES, SO YOU CAN GO HOME LATER THIS MORNING.
DO YOU HAVE ANY QUESTIONS?
WILL I... UM... FEEL ANYTHING?
YOU'LL BE UNDER TWILIGHT SEDATION FOR THE PROCEDURE. YOU WON'T FEEL ANY PAIN.
NOW, LET'S PUT THIS MASK ON.
IT'S JUST OXYGEN.
JOANNE, PLEASE COUNT BACKWARDS FROM TEN.
TEN...
NINE...
EIGHT...
SEVEN...
SIX...

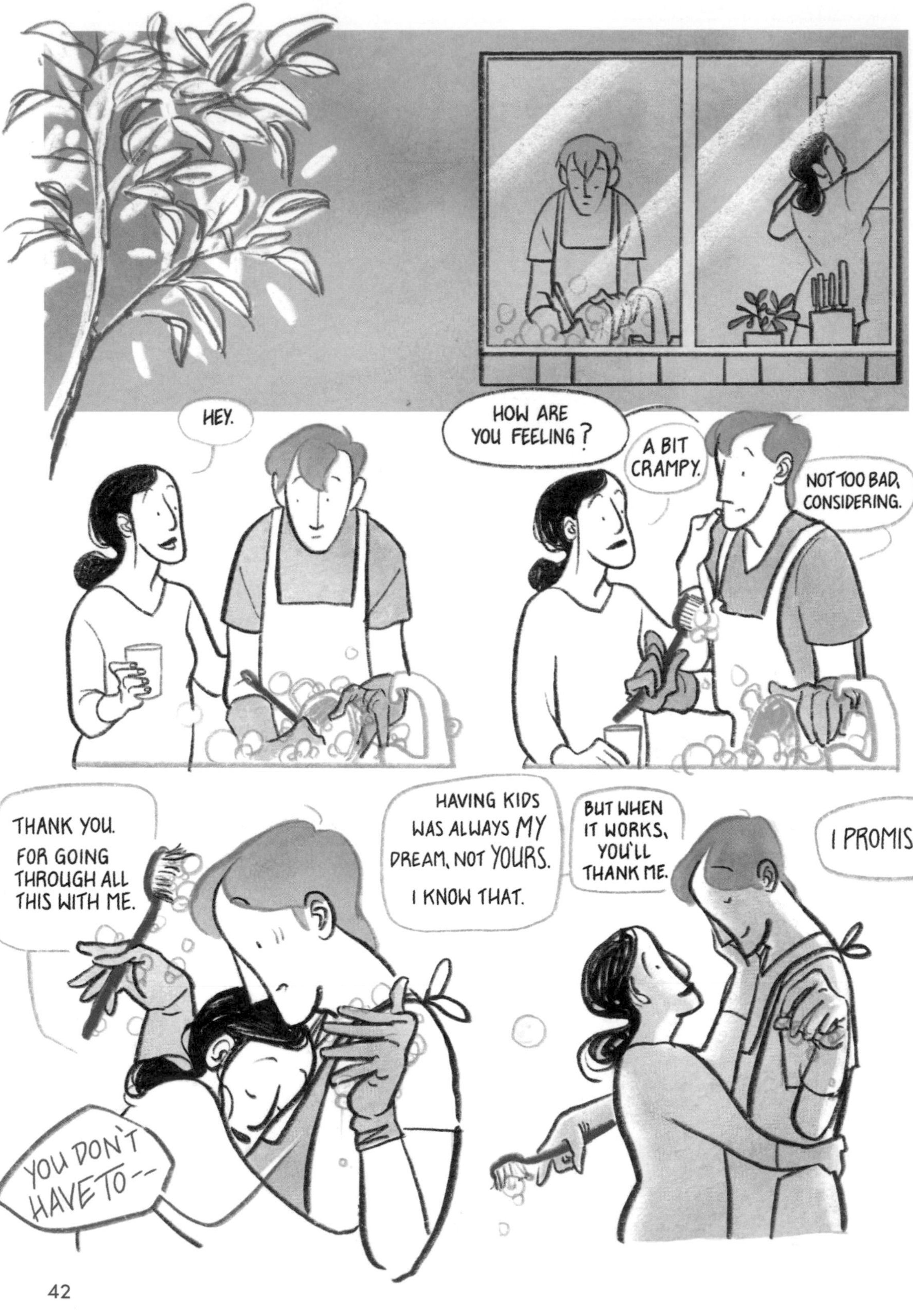
HEY.
HOW ARE YOU FEELING?
A BIT CRAMPY.
NOT TOO BAD, CONSIDERING.
THANK YOU. FOR GOING THROUGH ALL THIS WITH ME.
YOU DON'T HAVE TO--
HAVING KIDS WAS ALWAYS MY DREAM, NOT YOURS. I KNOW THAT.
BUT WHEN IT WORKS, YOU'LL THANK ME.
I PROMISE.

CHAPTER 3:

THE BIG DAY

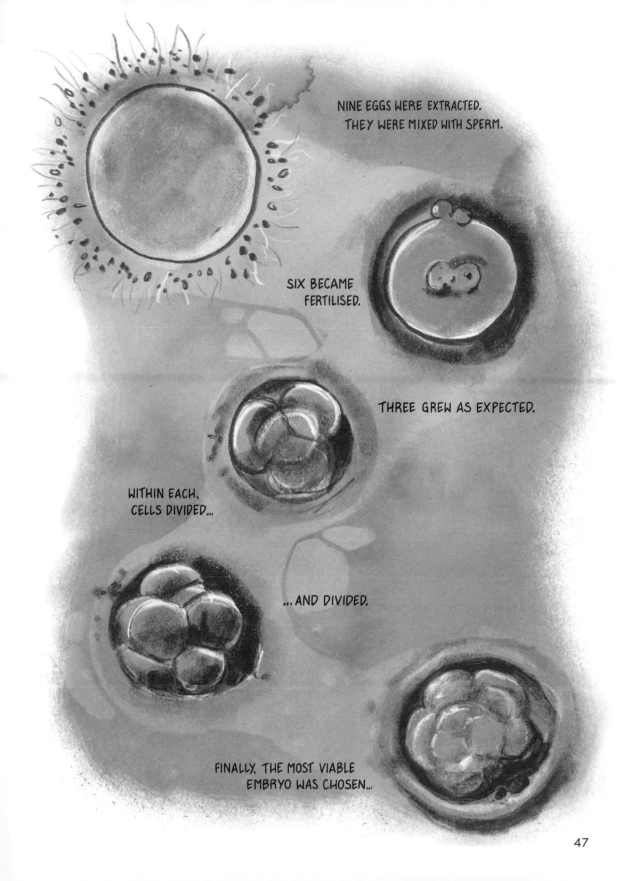

NINE EGGS WERE EXTRACTED.
THEY WERE MIXED WITH SPERM.

SIX BECAME
FERTILISED.

THREE GREW AS EXPECTED.

WITHIN EACH,
CELLS DIVIDED...

...AND DIVIDED.

FINALLY, THE MOST VIABLE
EMBRYO WAS CHOSEN...

... FOR TRANSFER.

NERVOUS.

EXCITED.

SO, HOW ARE WE FEELING TODAY?

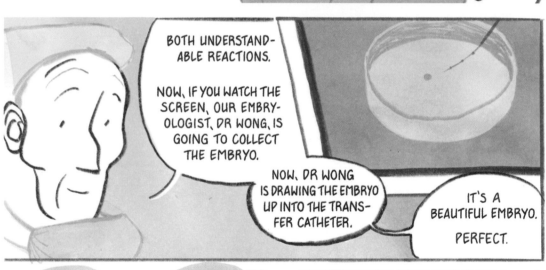

BOTH UNDERSTAND-ABLE REACTIONS.

NOW, IF YOU WATCH THE SCREEN, OUR EMBRY-OLOGIST, DR WONG, IS GOING TO COLLECT THE EMBRYO.

NOW, DR WONG IS DRAWING THE EMBRYO UP INTO THE TRANS-FER CATHETER.

IT'S A BEAUTIFUL EMBRYO.

PERFECT.

READY?

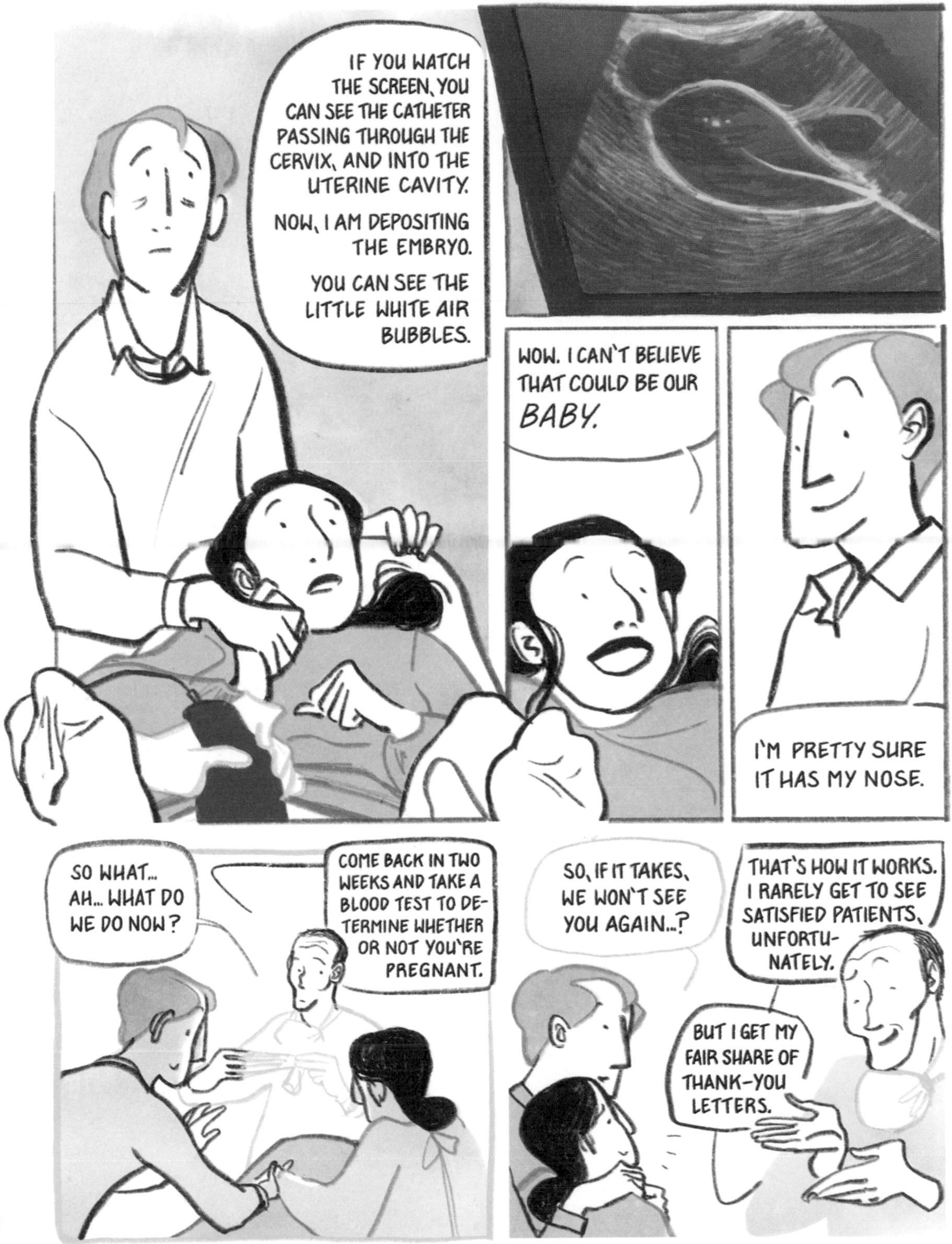
IF YOU WATCH THE SCREEN, YOU CAN SEE THE CATHETER PASSING THROUGH THE CERVIX, AND INTO THE UTERINE CAVITY.
NOW, I AM DEPOSITING THE EMBRYO.
YOU CAN SEE THE LITTLE WHITE AIR BUBBLES.
WOW. I CAN'T BELIEVE THAT COULD BE OUR *BABY.*
I'M PRETTY SURE IT HAS MY NOSE.
SO WHAT... AH... WHAT DO WE DO NOW?
COME BACK IN TWO WEEKS AND TAKE A BLOOD TEST TO DETERMINE WHETHER OR NOT YOU'RE PREGNANT.
SO, IF IT TAKES, WE WON'T SEE YOU AGAIN...?
THAT'S HOW IT WORKS. I RARELY GET TO SEE SATISFIED PATIENTS, UNFORTUNATELY.
BUT I GET MY FAIR SHARE OF THANK-YOU LETTERS.

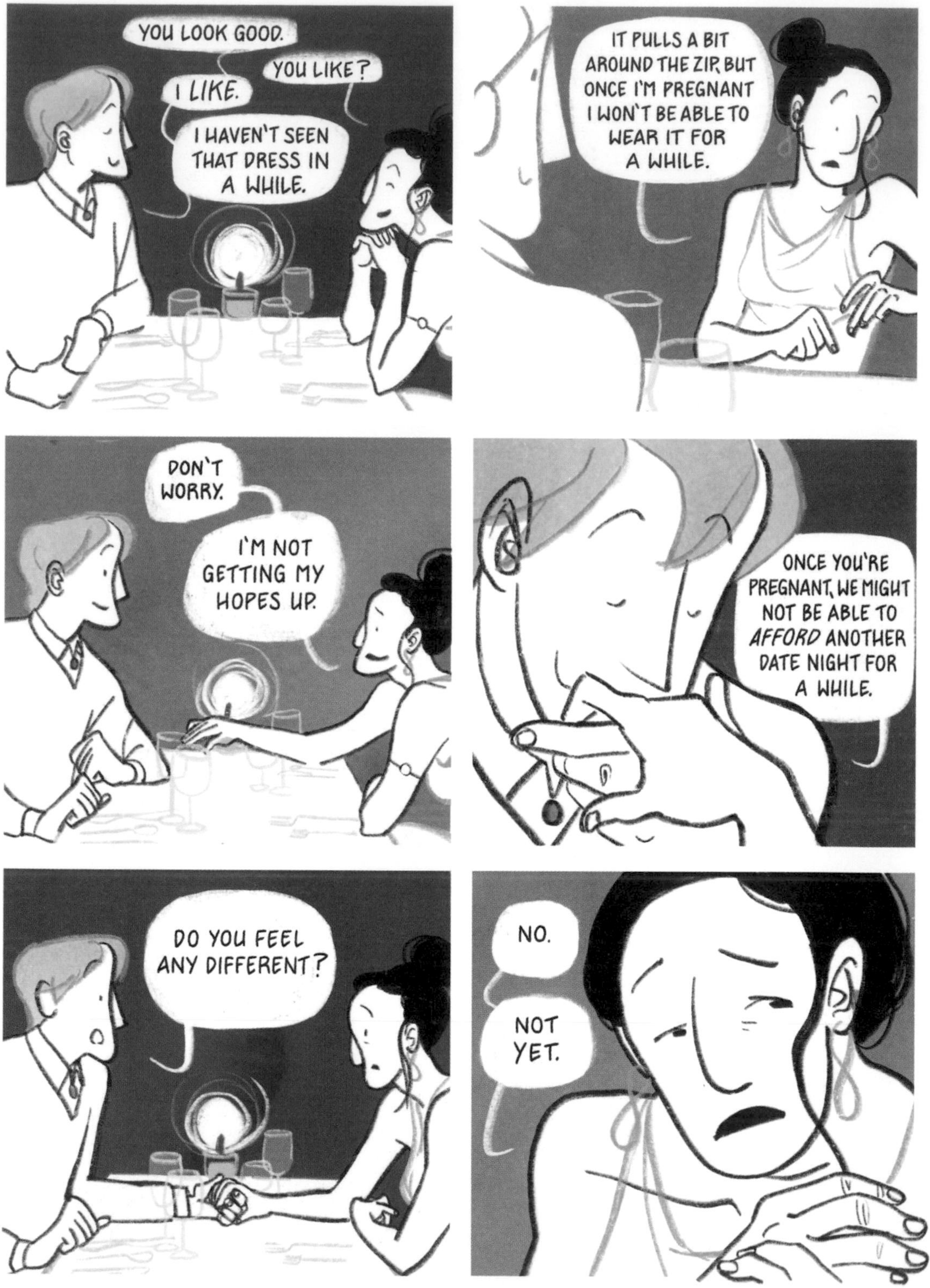
YOU LOOK GOOD.
YOU LIKE?
I LIKE.
I HAVEN'T SEEN THAT DRESS IN A WHILE.
IT PULLS A BIT AROUND THE ZIP, BUT ONCE I'M PREGNANT I WON'T BE ABLE TO WEAR IT FOR A WHILE.
DON'T WORRY.
I'M NOT GETTING MY HOPES UP.
ONCE YOU'RE PREGNANT, WE MIGHT NOT BE ABLE TO *AFFORD* ANOTHER DATE NIGHT FOR A WHILE.
DO YOU FEEL ANY DIFFERENT?
NO.
NOT YET.

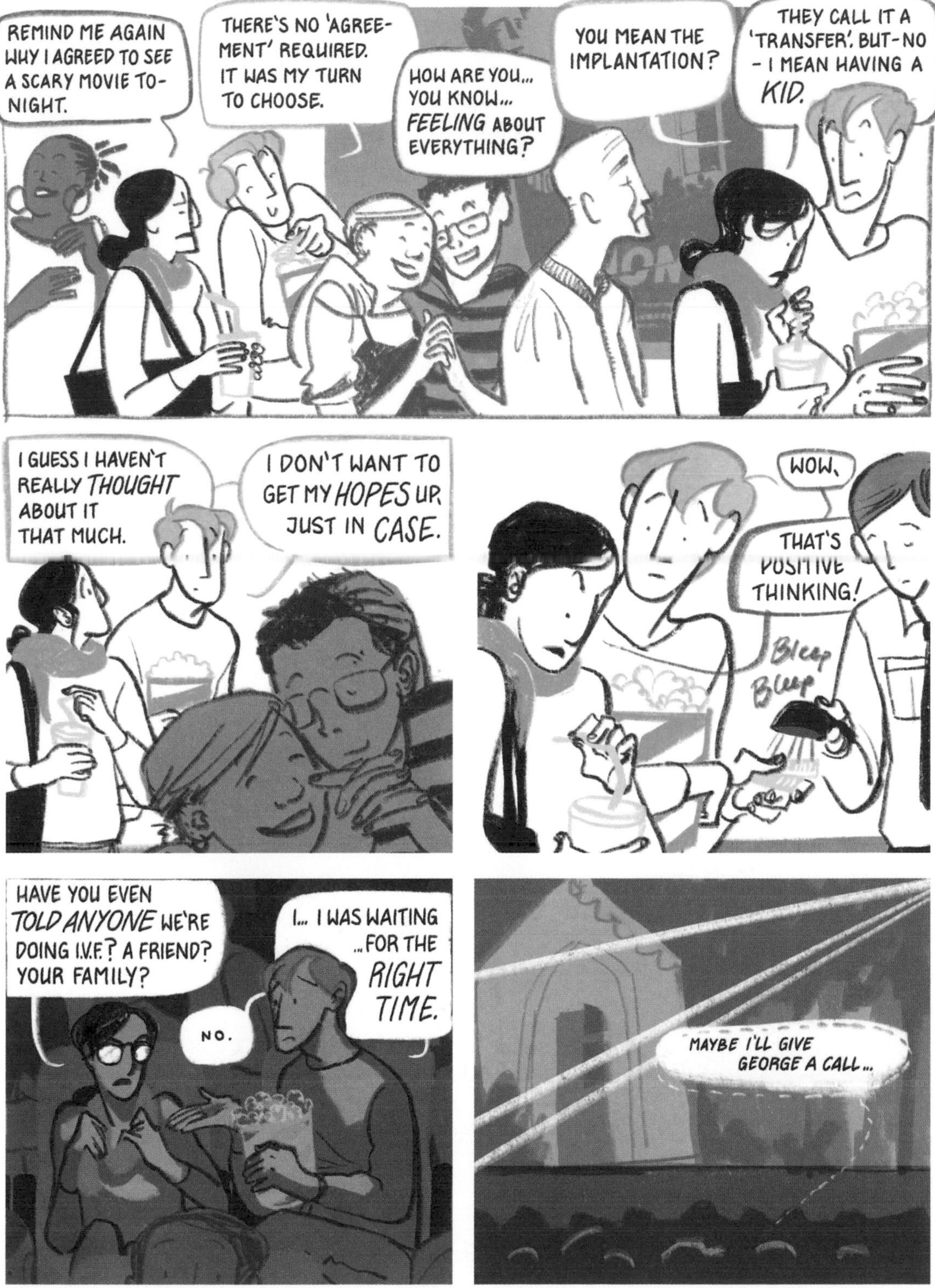
REMIND ME AGAIN WHY I AGREED TO SEE A SCARY MOVIE TO-NIGHT.
THERE'S NO 'AGREE-MENT' REQUIRED. IT WAS MY TURN TO CHOOSE.
HOW ARE YOU... YOU KNOW... FEELING ABOUT EVERYTHING?
YOU MEAN THE IMPLANTATION?
THEY CALL IT A 'TRANSFER'. BUT-NO - I MEAN HAVING A KID.
I GUESS I HAVEN'T REALLY THOUGHT ABOUT IT THAT MUCH.
I DON'T WANT TO GET MY HOPES UP, JUST IN CASE.
WOW,
THAT'S POSITIVE THINKING!
Bleep Bleep
HAVE YOU EVEN TOLD ANYONE WE'RE DOING I.V.F.? A FRIEND? YOUR FAMILY?
NO.
I... I WAS WAITING ...FOR THE RIGHT TIME.
MAYBE I'LL GIVE GEORGE A CALL...

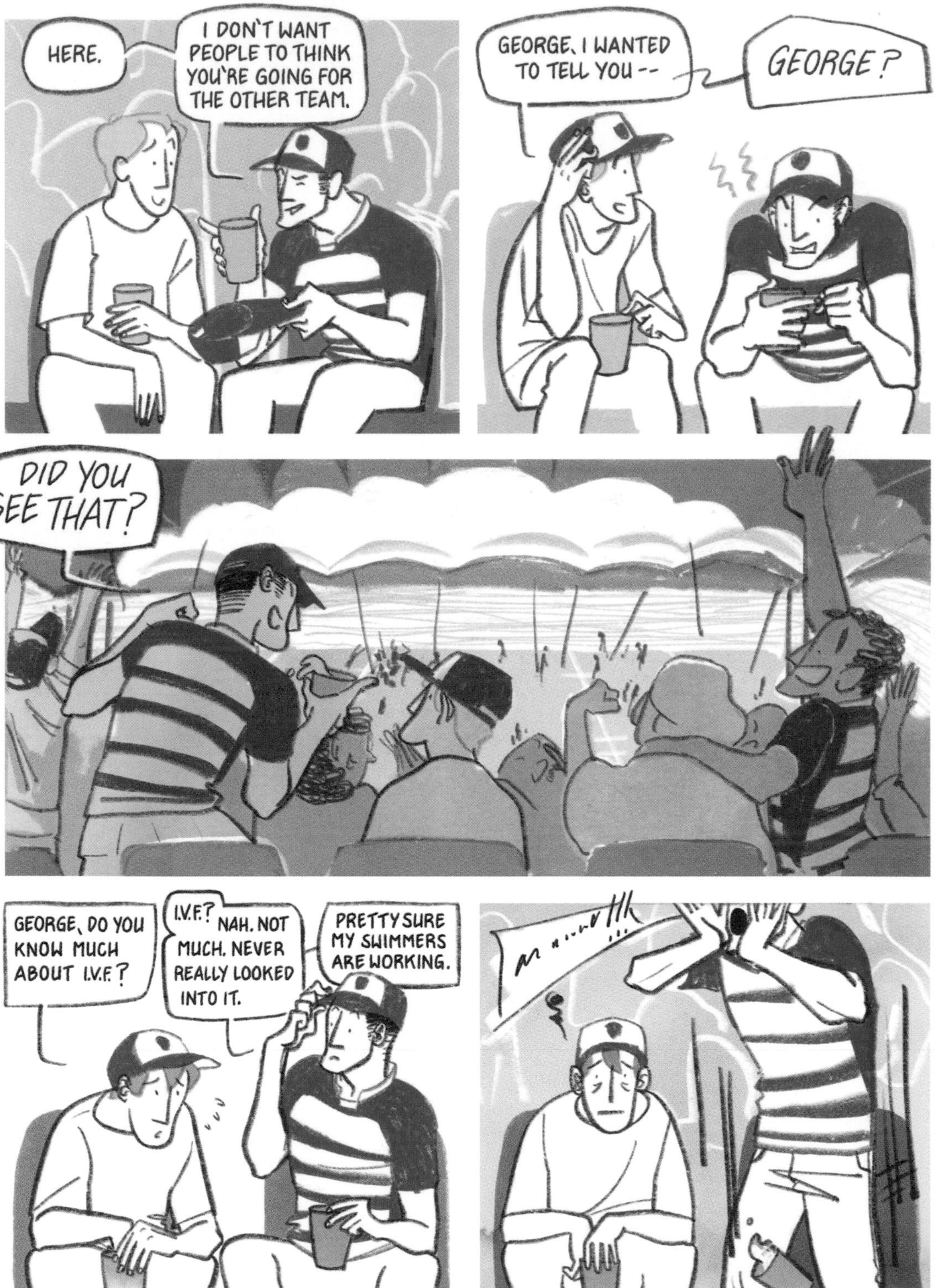
HERE.
I DON'T WANT PEOPLE TO THINK YOU'RE GOING FOR THE OTHER TEAM.
GEORGE, I WANTED TO TELL YOU --
GEORGE?
DID YOU SEE THAT?
GEORGE, DO YOU KNOW MUCH ABOUT I.V.F?
I.V.F? NAH. NOT MUCH. NEVER REALLY LOOKED INTO IT.
PRETTY SURE MY SWIMMERS ARE WORKING.

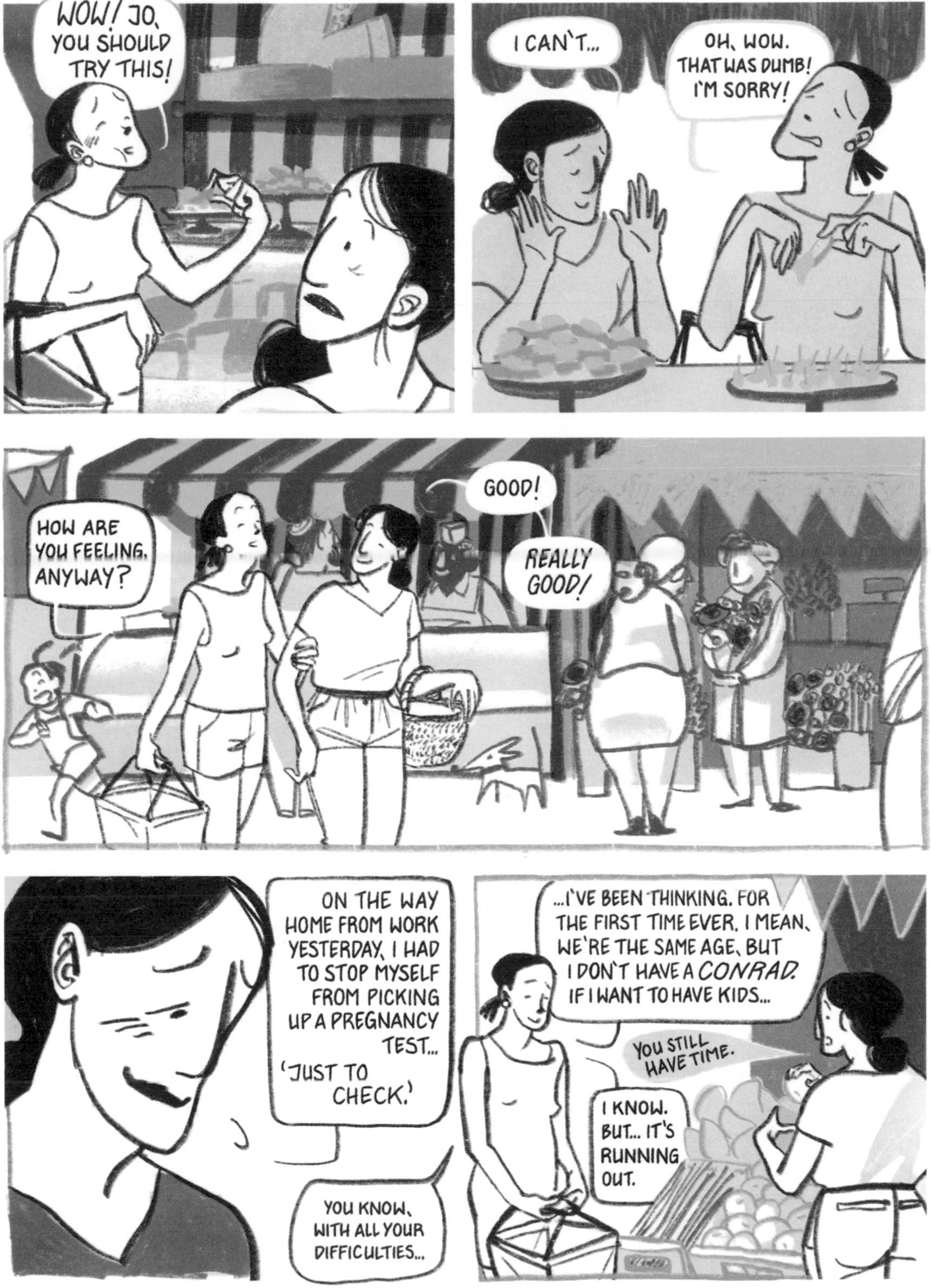
WOW! JO, YOU SHOULD TRY THIS!
I CAN'T...
OH, WOW. THAT WAS DUMB! I'M SORRY!
HOW ARE YOU FEELING, ANYWAY?
GOOD!
REALLY GOOD!
ON THE WAY HOME FROM WORK YESTERDAY, I HAD TO STOP MYSELF FROM PICKING UP A PREGNANCY TEST...
'JUST TO CHECK.'
YOU KNOW, WITH ALL YOUR DIFFICULTIES...
...I'VE BEEN THINKING. FOR THE FIRST TIME EVER, I MEAN. WE'RE THE SAME AGE, BUT I DON'T HAVE A CONRAD. IF I WANT TO HAVE KIDS...
YOU STILL HAVE TIME.
I KNOW. BUT... IT'S RUNNING OUT.

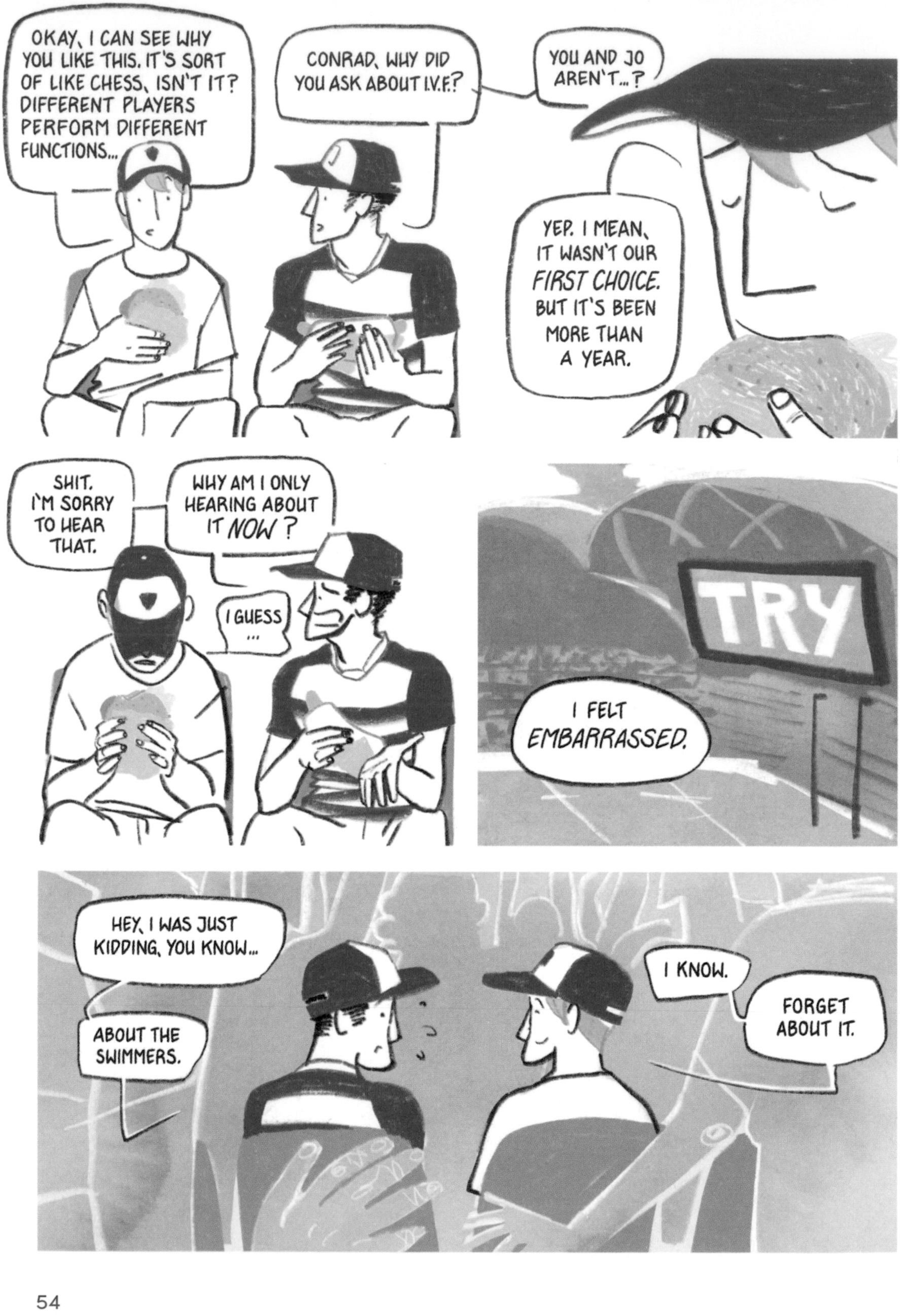
OKAY, I CAN SEE WHY YOU LIKE THIS. IT'S SORT OF LIKE CHESS, ISN'T IT? DIFFERENT PLAYERS PERFORM DIFFERENT FUNCTIONS...
CONRAD, WHY DID YOU ASK ABOUT I.V.F.?
YOU AND JO AREN'T...?
YEP. I MEAN, IT WASN'T OUR *FIRST CHOICE.* BUT IT'S BEEN MORE THAN A YEAR.
SHIT. I'M SORRY TO HEAR THAT.
WHY AM I ONLY HEARING ABOUT IT *NOW*?
I GUESS ...
TRY
I FELT *EMBARRASSED.*
HEY, I WAS JUST KIDDING, YOU KNOW...
ABOUT THE SWIMMERS.
I KNOW.
FORGET ABOUT IT.

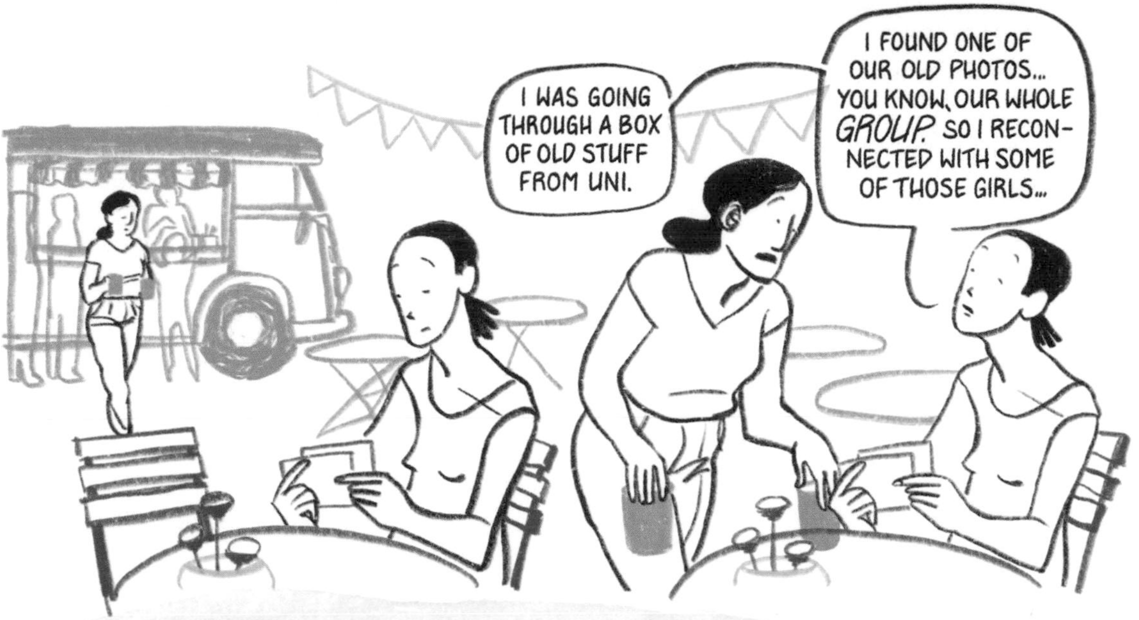
I WAS GOING THROUGH A BOX OF OLD STUFF FROM UNI.
I FOUND ONE OF OUR OLD PHOTOS... YOU KNOW, OUR WHOLE *GROUP*. SO I RECONNECTED WITH SOME OF THOSE GIRLS...

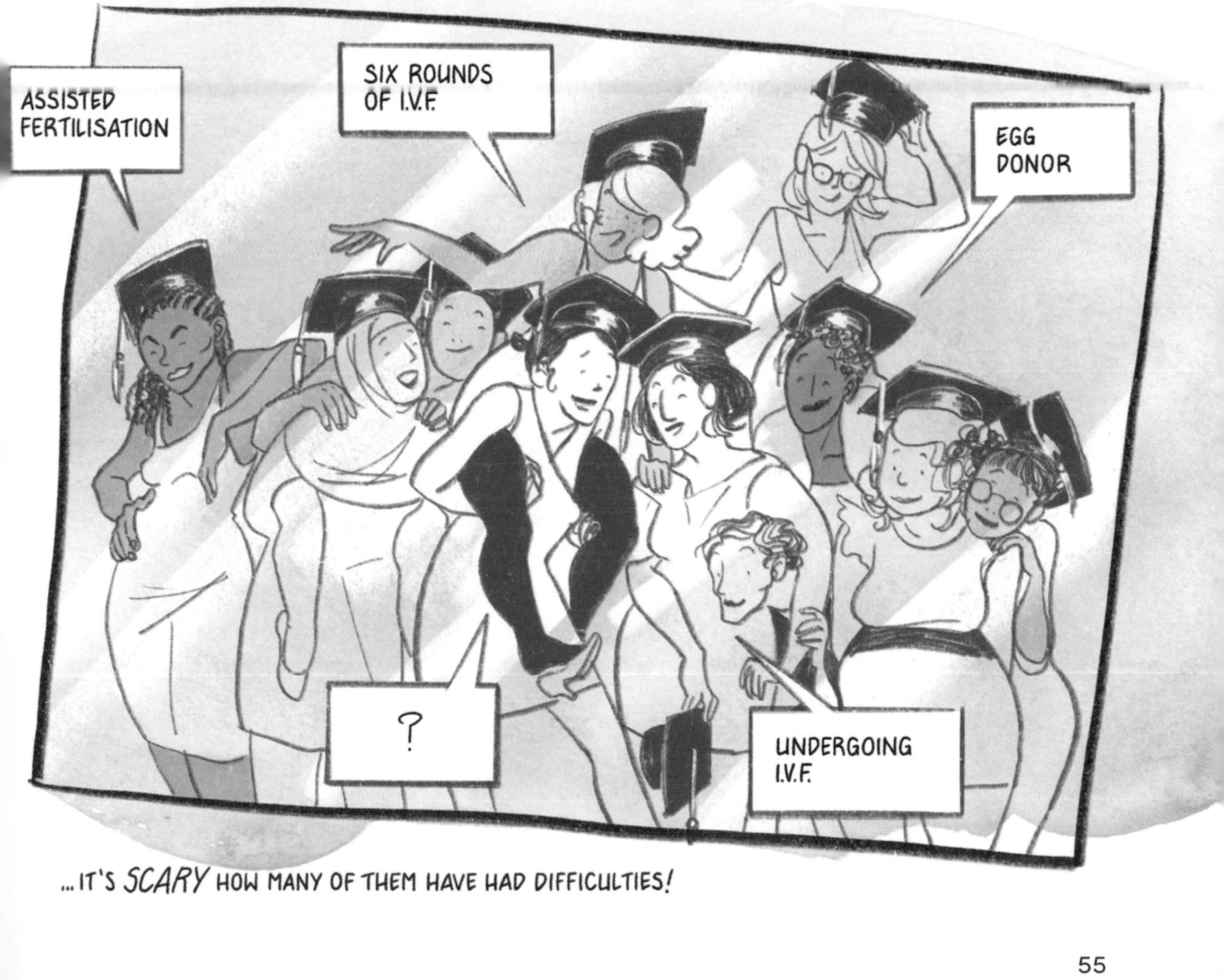
ASSISTED FERTILISATION
SIX ROUNDS OF I.V.F.
EGG DONOR
?
UNDERGOING I.V.F.

...IT'S *SCARY* HOW MANY OF THEM HAVE HAD DIFFICULTIES!

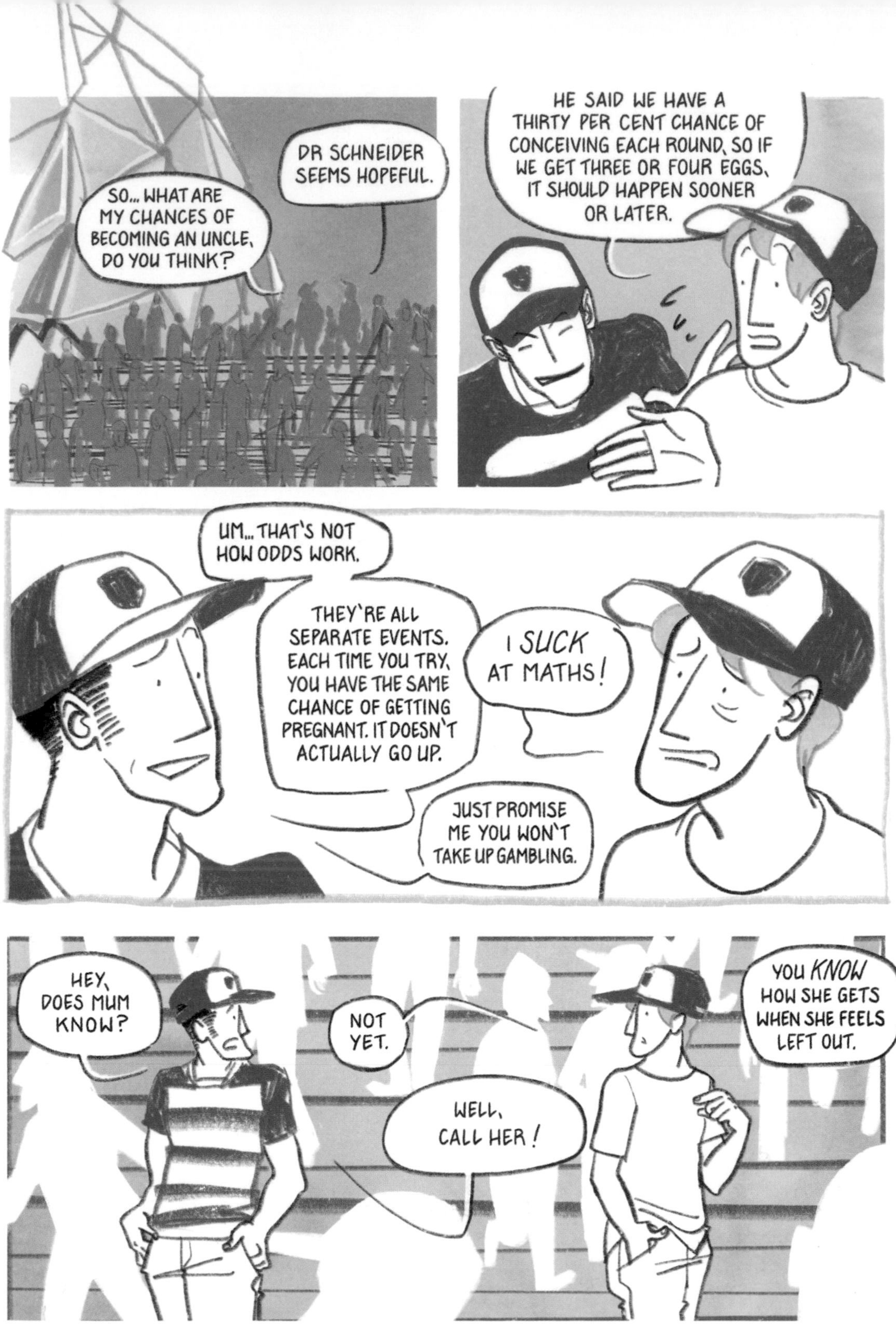
SO... WHAT ARE MY CHANCES OF BECOMING AN UNCLE, DO YOU THINK?
DR SCHNEIDER SEEMS HOPEFUL.
HE SAID WE HAVE A THIRTY PER CENT CHANCE OF CONCEIVING EACH ROUND, SO IF WE GET THREE OR FOUR EGGS, IT SHOULD HAPPEN SOONER OR LATER.
UM... THAT'S NOT HOW ODDS WORK.
THEY'RE ALL SEPARATE EVENTS. EACH TIME YOU TRY, YOU HAVE THE SAME CHANCE OF GETTING PREGNANT. IT DOESN'T ACTUALLY GO UP.
I SUCK AT MATHS!
JUST PROMISE ME YOU WON'T TAKE UP GAMBLING.
HEY, DOES MUM KNOW?
NOT YET.
YOU KNOW HOW SHE GETS WHEN SHE FEELS LEFT OUT.
WELL, CALL HER!

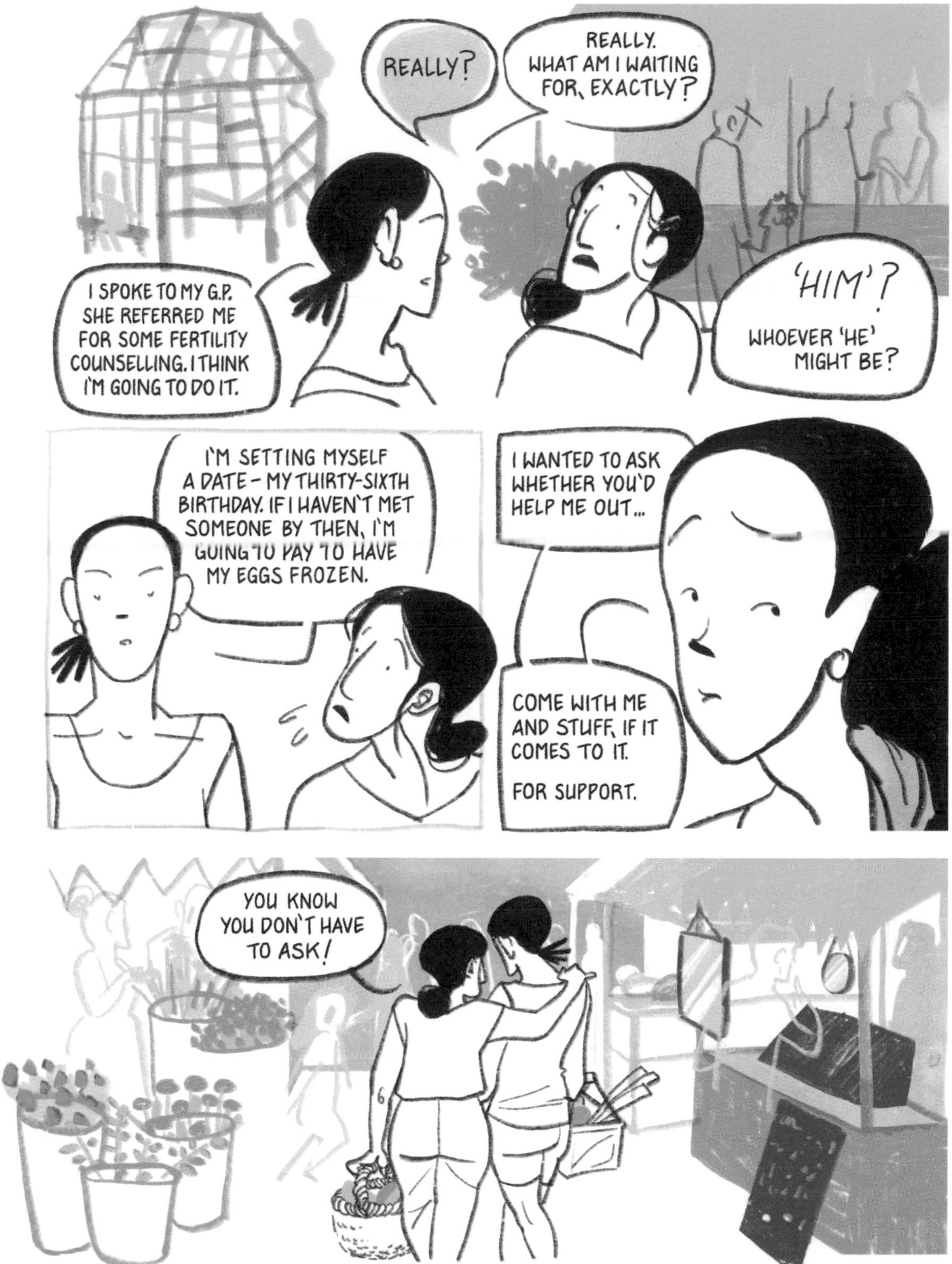
REALLY?
REALLY. WHAT AM I WAITING FOR, EXACTLY?
I SPOKE TO MY G.P. SHE REFERRED ME FOR SOME FERTILITY COUNSELLING. I THINK I'M GOING TO DO IT.
'HIM'? WHOEVER 'HE' MIGHT BE?
I'M SETTING MYSELF A DATE – MY THIRTY-SIXTH BIRTHDAY. IF I HAVEN'T MET SOMEONE BY THEN, I'M GOING TO PAY TO HAVE MY EGGS FROZEN.
I WANTED TO ASK WHETHER YOU'D HELP ME OUT...
COME WITH ME AND STUFF, IF IT COMES TO IT.
FOR SUPPORT.
YOU KNOW YOU DON'T HAVE TO ASK!

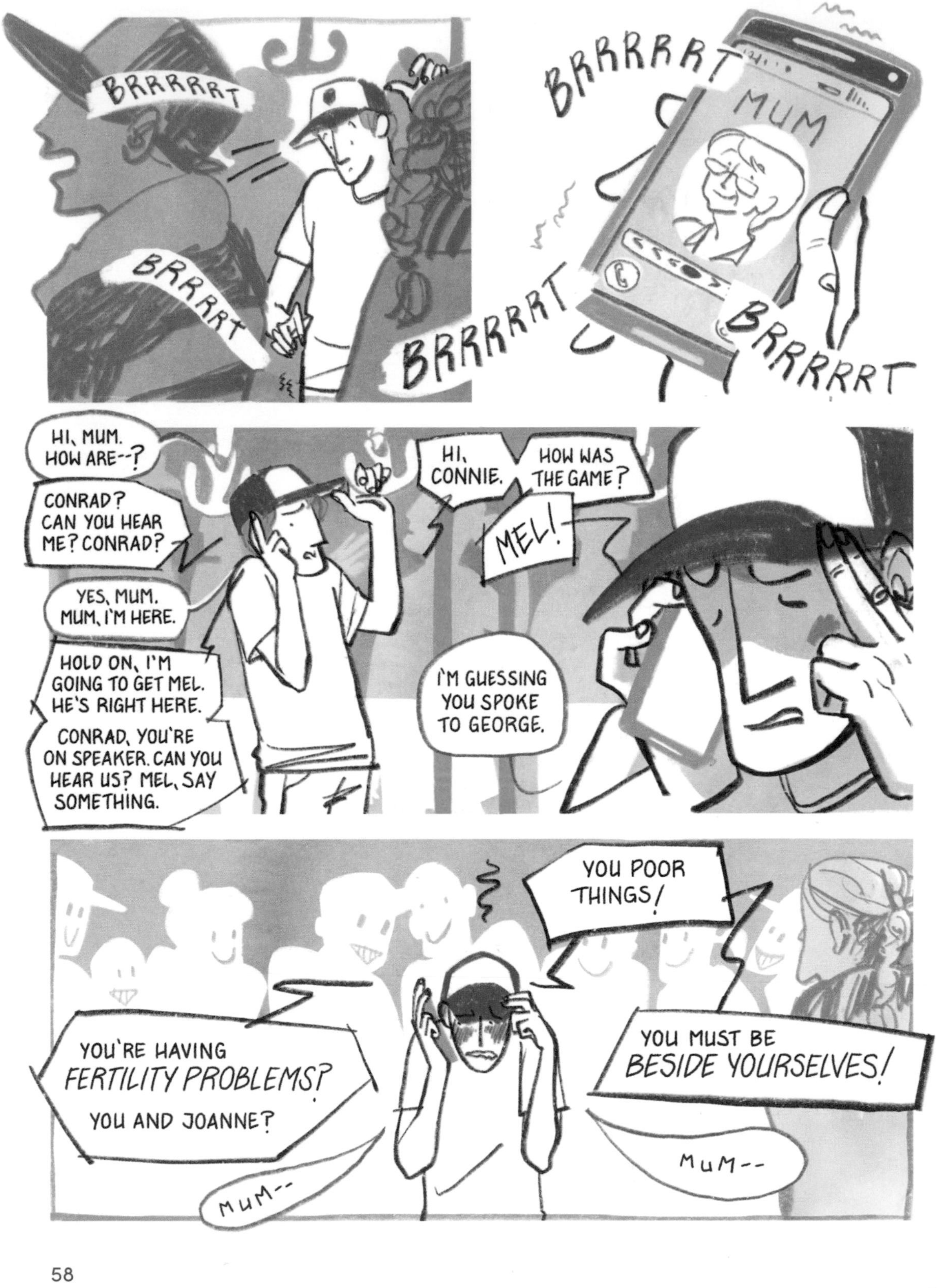
BRRRRRT
BRRRRT
BRRRRRT
MUM
BRRRRRT
BRRRRRT
HI, MUM. HOW ARE--?
CONRAD? CAN YOU HEAR ME? CONRAD?
YES, MUM. MUM, I'M HERE.
HOLD ON, I'M GOING TO GET MEL. HE'S RIGHT HERE.
CONRAD, YOU'RE ON SPEAKER. CAN YOU HEAR US? MEL, SAY SOMETHING.
HI, CONNIE.
HOW WAS THE GAME?
MEL!
I'M GUESSING YOU SPOKE TO GEORGE.
YOU POOR THINGS!
YOU'RE HAVING FERTILITY PROBLEMS? YOU AND JOANNE?
YOU MUST BE BESIDE YOURSELVES!
MUM--
MUM--

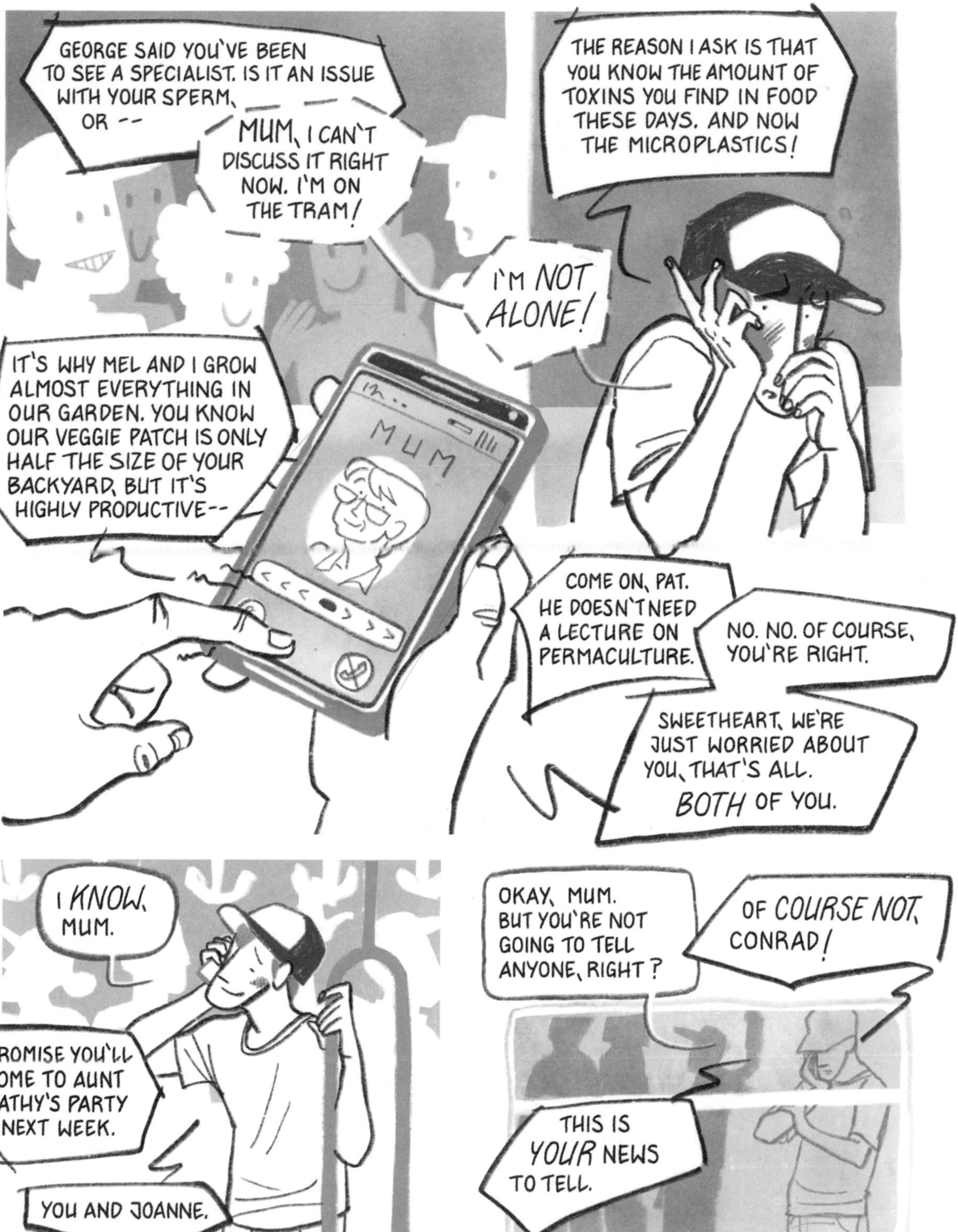
GEORGE SAID YOU'VE BEEN TO SEE A SPECIALIST. IS IT AN ISSUE WITH YOUR SPERM, OR --
MUM, I CAN'T DISCUSS IT RIGHT NOW. I'M ON THE TRAM!
THE REASON I ASK IS THAT YOU KNOW THE AMOUNT OF TOXINS YOU FIND IN FOOD THESE DAYS. AND NOW THE MICROPLASTICS!
I'M NOT ALONE!
IT'S WHY MEL AND I GROW ALMOST EVERYTHING IN OUR GARDEN. YOU KNOW OUR VEGGIE PATCH IS ONLY HALF THE SIZE OF YOUR BACKYARD, BUT IT'S HIGHLY PRODUCTIVE--
MUM
COME ON, PAT. HE DOESN'T NEED A LECTURE ON PERMACULTURE.
NO. NO. OF COURSE, YOU'RE RIGHT.
SWEETHEART, WE'RE JUST WORRIED ABOUT YOU, THAT'S ALL. BOTH OF YOU.
I KNOW, MUM.
PROMISE YOU'LL COME TO AUNT CATHY'S PARTY NEXT WEEK.
YOU AND JOANNE.
OKAY, MUM. BUT YOU'RE NOT GOING TO TELL ANYONE, RIGHT?
OF COURSE NOT, CONRAD!
THIS IS YOUR NEWS TO TELL.

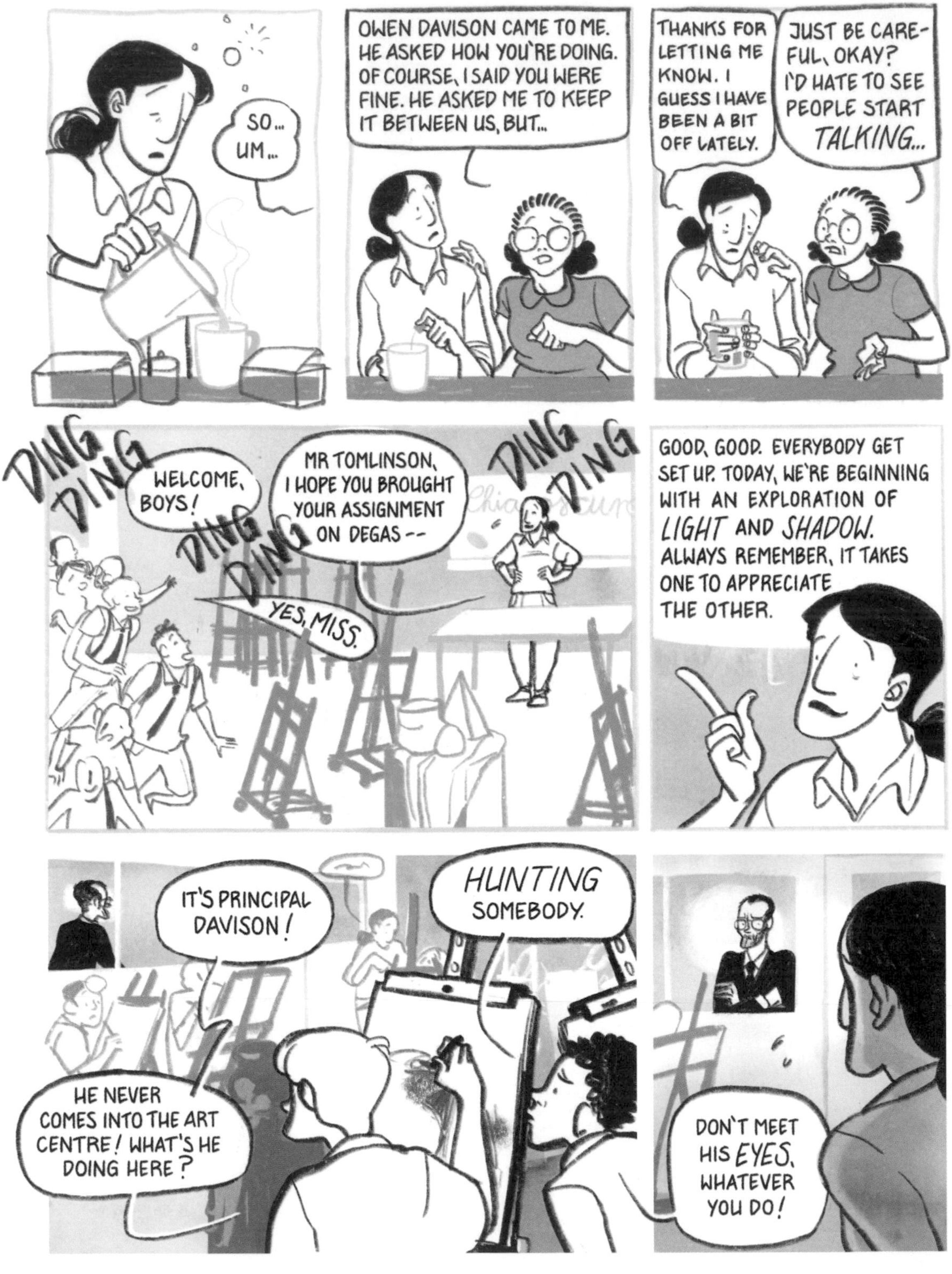

SO... UM...
OWEN DAVISON CAME TO ME. HE ASKED HOW YOU'RE DOING. OF COURSE, I SAID YOU WERE FINE. HE ASKED ME TO KEEP IT BETWEEN US, BUT...
THANKS FOR LETTING ME KNOW. I GUESS I HAVE BEEN A BIT OFF LATELY.
JUST BE CAREFUL, OKAY? I'D HATE TO SEE PEOPLE START TALKING...
DING DING
WELCOME, BOYS!
DING DING
MR TOMLINSON, I HOPE YOU BROUGHT YOUR ASSIGNMENT ON DEGAS --
DING DING
YES, MISS.
GOOD, GOOD. EVERYBODY GET SET UP. TODAY, WE'RE BEGINNING WITH AN EXPLORATION OF LIGHT AND SHADOW. ALWAYS REMEMBER, IT TAKES ONE TO APPRECIATE THE OTHER.
IT'S PRINCIPAL DAVISON!
HUNTING SOMEBODY.
HE NEVER COMES INTO THE ART CENTRE! WHAT'S HE DOING HERE?
DON'T MEET HIS EYES, WHATEVER YOU DO!

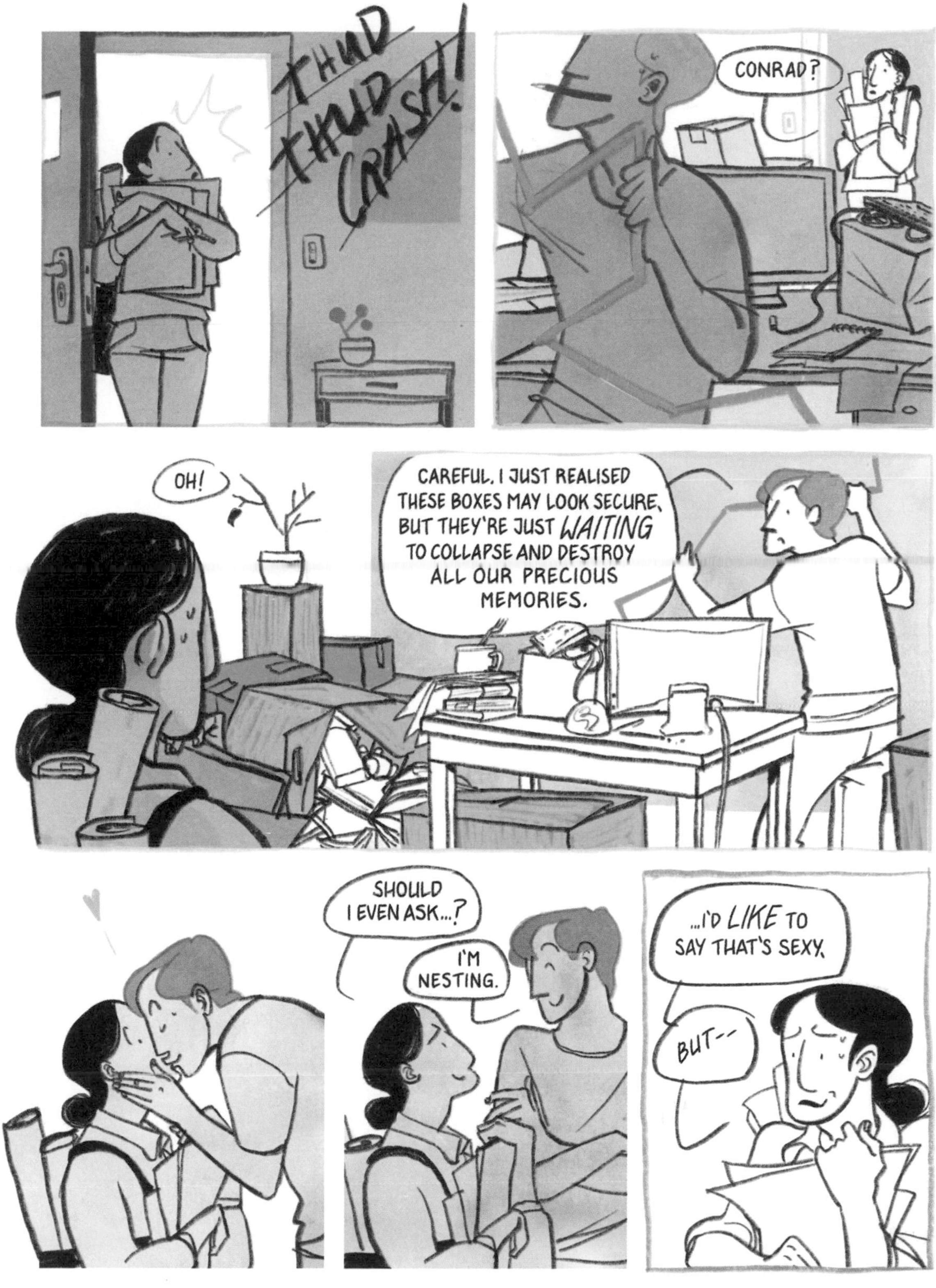
THUD
THUD
CRASH!
CONRAD?
OH!
CAREFUL. I JUST REALISED THESE BOXES MAY LOOK SECURE, BUT THEY'RE JUST WAITING TO COLLAPSE AND DESTROY ALL OUR PRECIOUS MEMORIES.
SHOULD I EVEN ASK...?
I'M NESTING.
...I'D LIKE TO SAY THAT'S SEXY,
BUT--

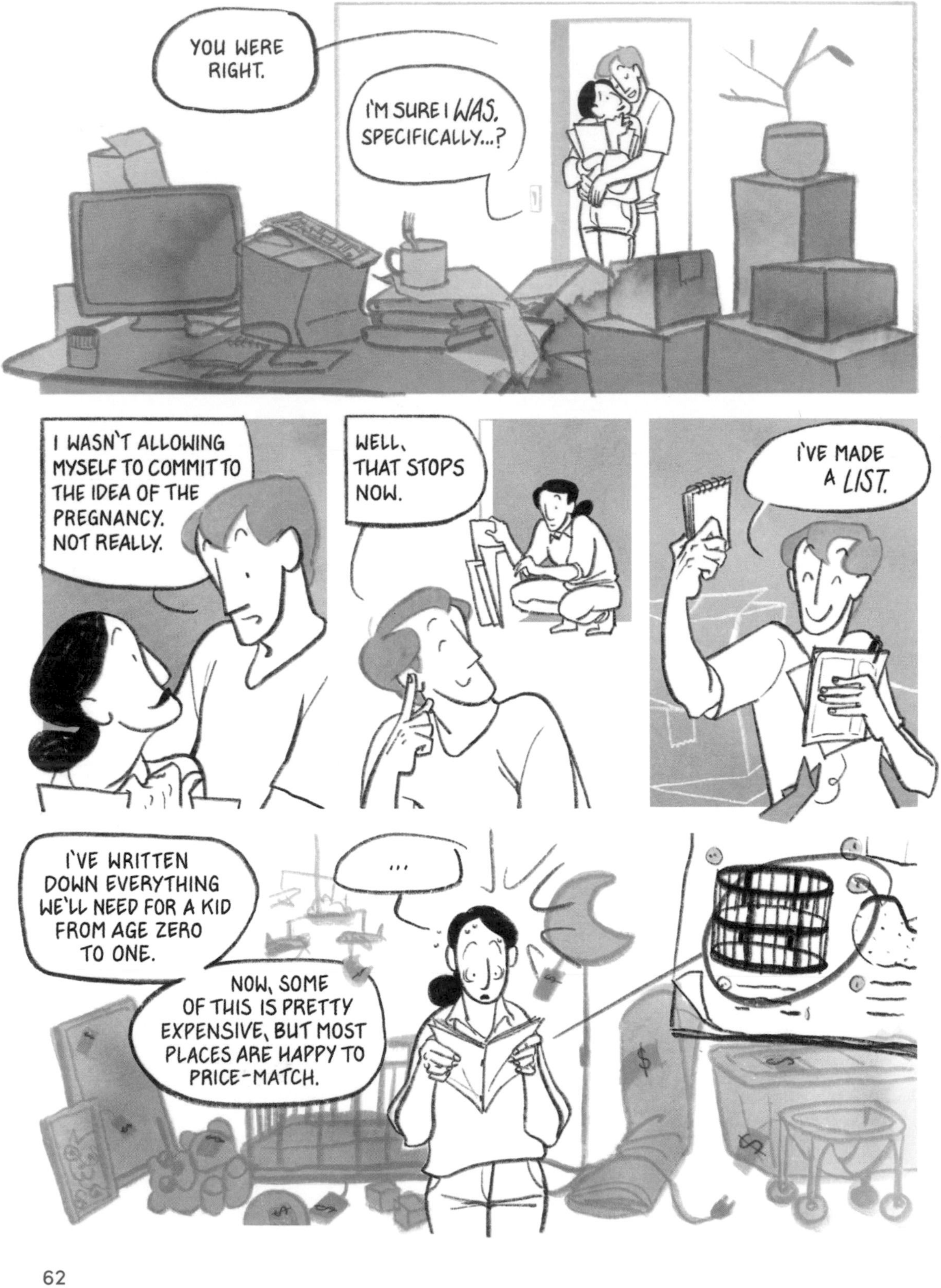
YOU WERE RIGHT.
I'M SURE I WAS. SPECIFICALLY...?
I WASN'T ALLOWING MYSELF TO COMMIT TO THE IDEA OF THE PREGNANCY. NOT REALLY.
WELL, THAT STOPS NOW.
I'VE MADE A LIST.
I'VE WRITTEN DOWN EVERYTHING WE'LL NEED FOR A KID FROM AGE ZERO TO ONE.
NOW, SOME OF THIS IS PRETTY EXPENSIVE, BUT MOST PLACES ARE HAPPY TO PRICE-MATCH.
...

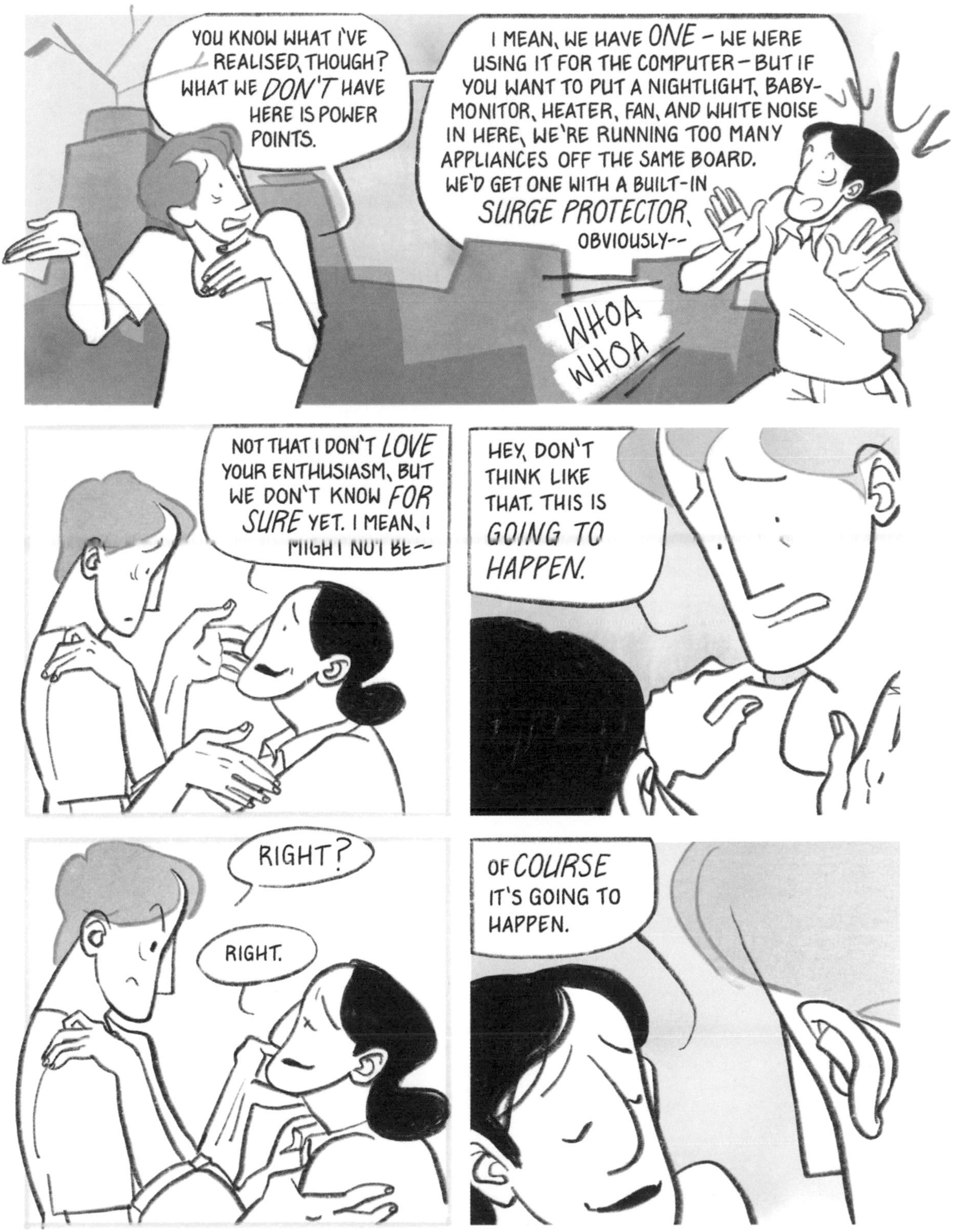
YOU KNOW WHAT I'VE REALISED, THOUGH? WHAT WE DON'T HAVE HERE IS POWER POINTS.
I MEAN, WE HAVE ONE - WE WERE USING IT FOR THE COMPUTER - BUT IF YOU WANT TO PUT A NIGHTLIGHT, BABY-MONITOR, HEATER, FAN, AND WHITE NOISE IN HERE, WE'RE RUNNING TOO MANY APPLIANCES OFF THE SAME BOARD. WE'D GET ONE WITH A BUILT-IN SURGE PROTECTOR, OBVIOUSLY--
WHOA WHOA
NOT THAT I DON'T LOVE YOUR ENTHUSIASM, BUT WE DON'T KNOW FOR SURE YET. I MEAN, I MIGHT NOT BE--
HEY, DON'T THINK LIKE THAT. THIS IS GOING TO HAPPEN.
RIGHT?
RIGHT.
OF COURSE IT'S GOING TO HAPPEN.

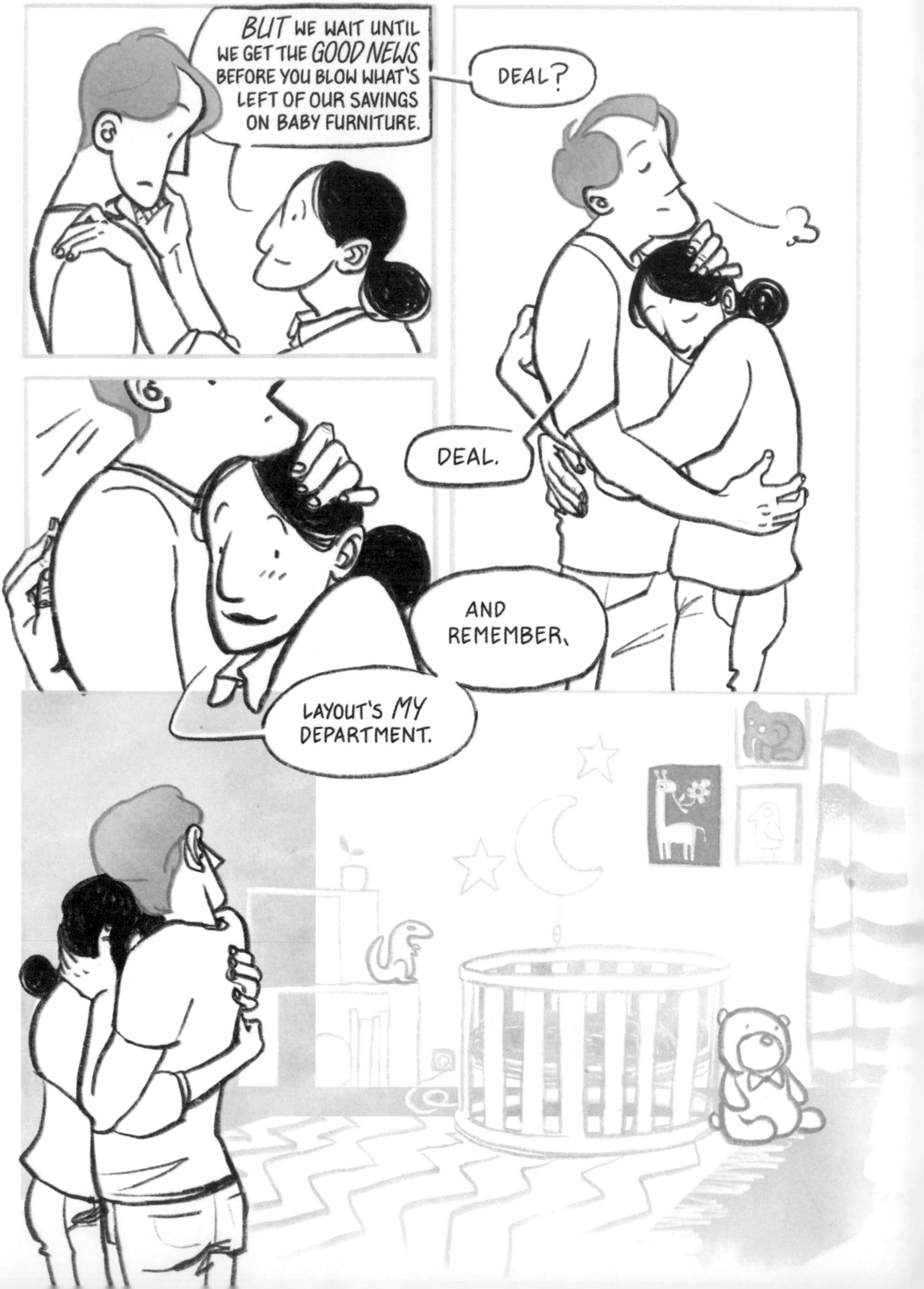
BUT WE WAIT UNTIL WE GET THE GOOD NEWS BEFORE YOU BLOW WHAT'S LEFT OF OUR SAVINGS ON BABY FURNITURE.
DEAL?
DEAL.
AND REMEMBER,
LAYOUT'S MY DEPARTMENT.

CHAPTER 4:

THE TEST

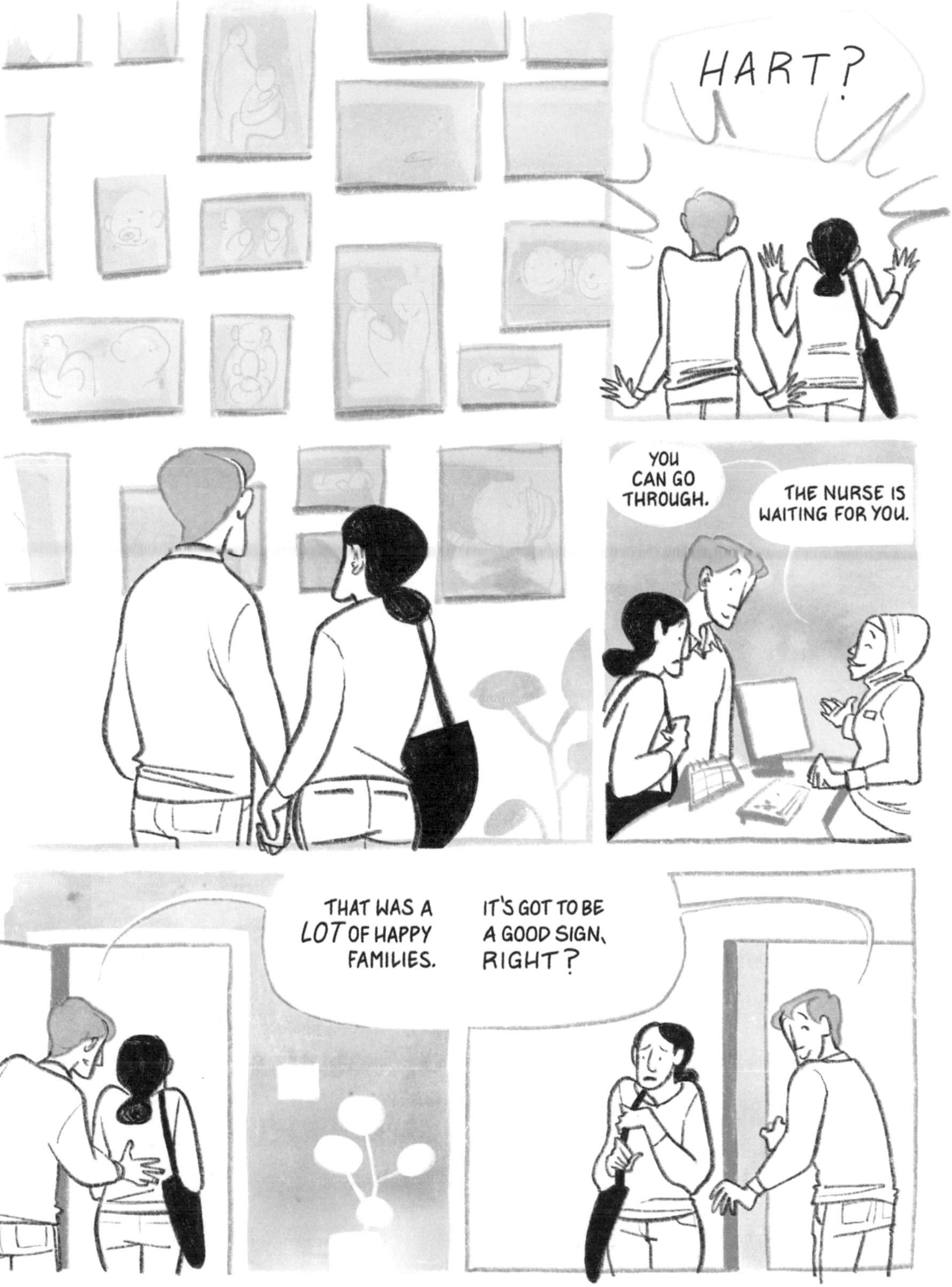
HART?
YOU CAN GO THROUGH.
THE NURSE IS WAITING FOR YOU.
THAT WAS A LOT OF HAPPY FAMILIES.
IT'S GOT TO BE A GOOD SIGN, RIGHT?

YOUR FILE SAYS THIS IS YOUR FIRST BLOOD TEST HERE.

YEAH.

FIRST AND LAST, WE HOPE.

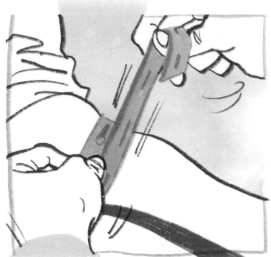

GOOD, STRONG VEINS.

THAT'S WHAT WE LIKE TO SEE.

THERE.

NOT TOO BAD, WAS IT?

EASY.

SO WE JUST WAIT...?

A COUPLE OF HOURS, SOME-BODY WILL BE IN TOUCH.

TRY TO RELAX.

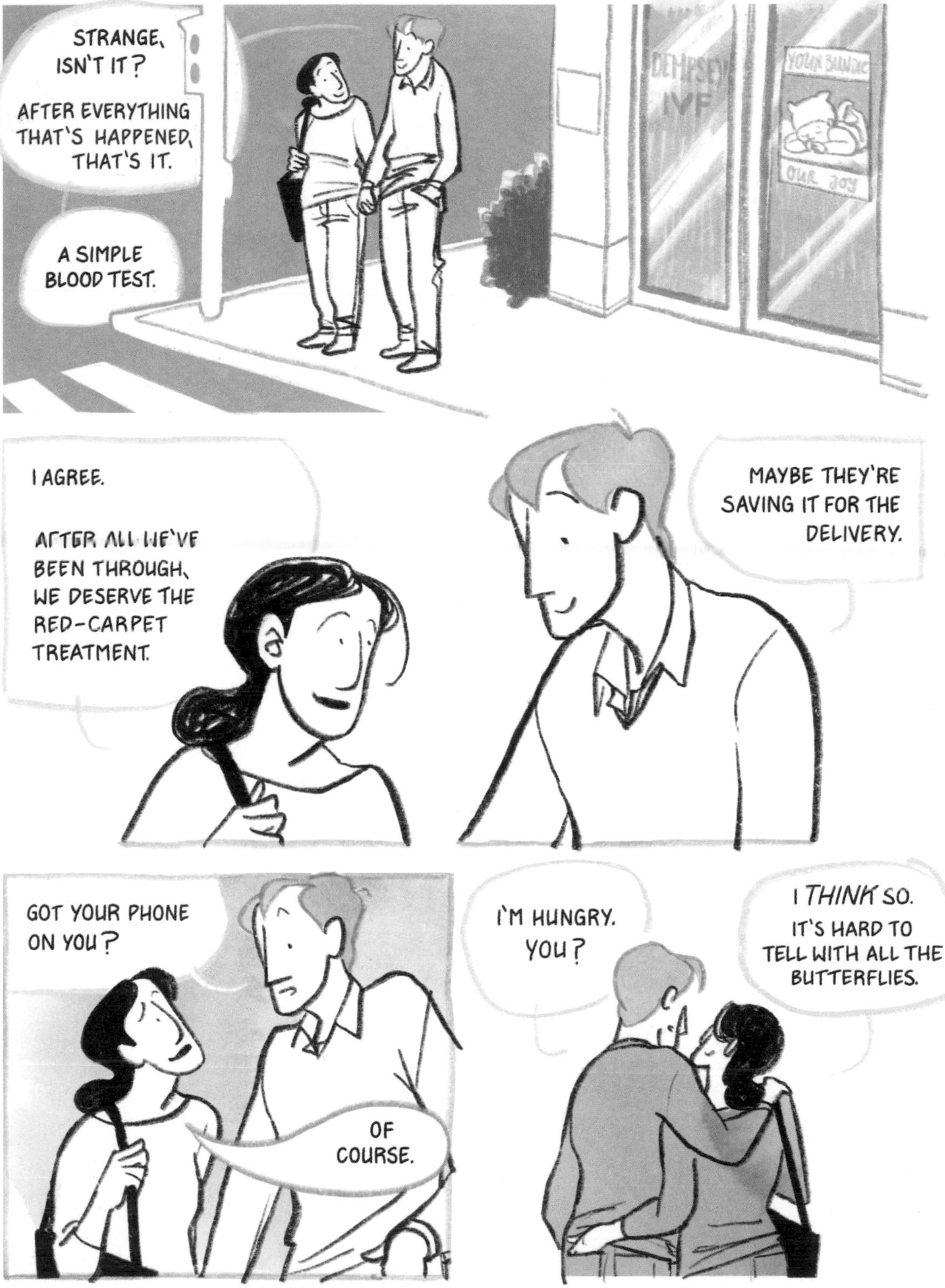
STRANGE, ISN'T IT?
AFTER EVERYTHING THAT'S HAPPENED, THAT'S IT.
A SIMPLE BLOOD TEST.
DEMPSEY IVF
YOUR BUNDLE
OUR JOY
I AGREE.
AFTER ALL WE'VE BEEN THROUGH, WE DESERVE THE RED-CARPET TREATMENT.
MAYBE THEY'RE SAVING IT FOR THE DELIVERY.
GOT YOUR PHONE ON YOU?
OF COURSE.
I'M HUNGRY. YOU?
I *THINK* SO.
IT'S HARD TO TELL WITH ALL THE BUTTERFLIES.

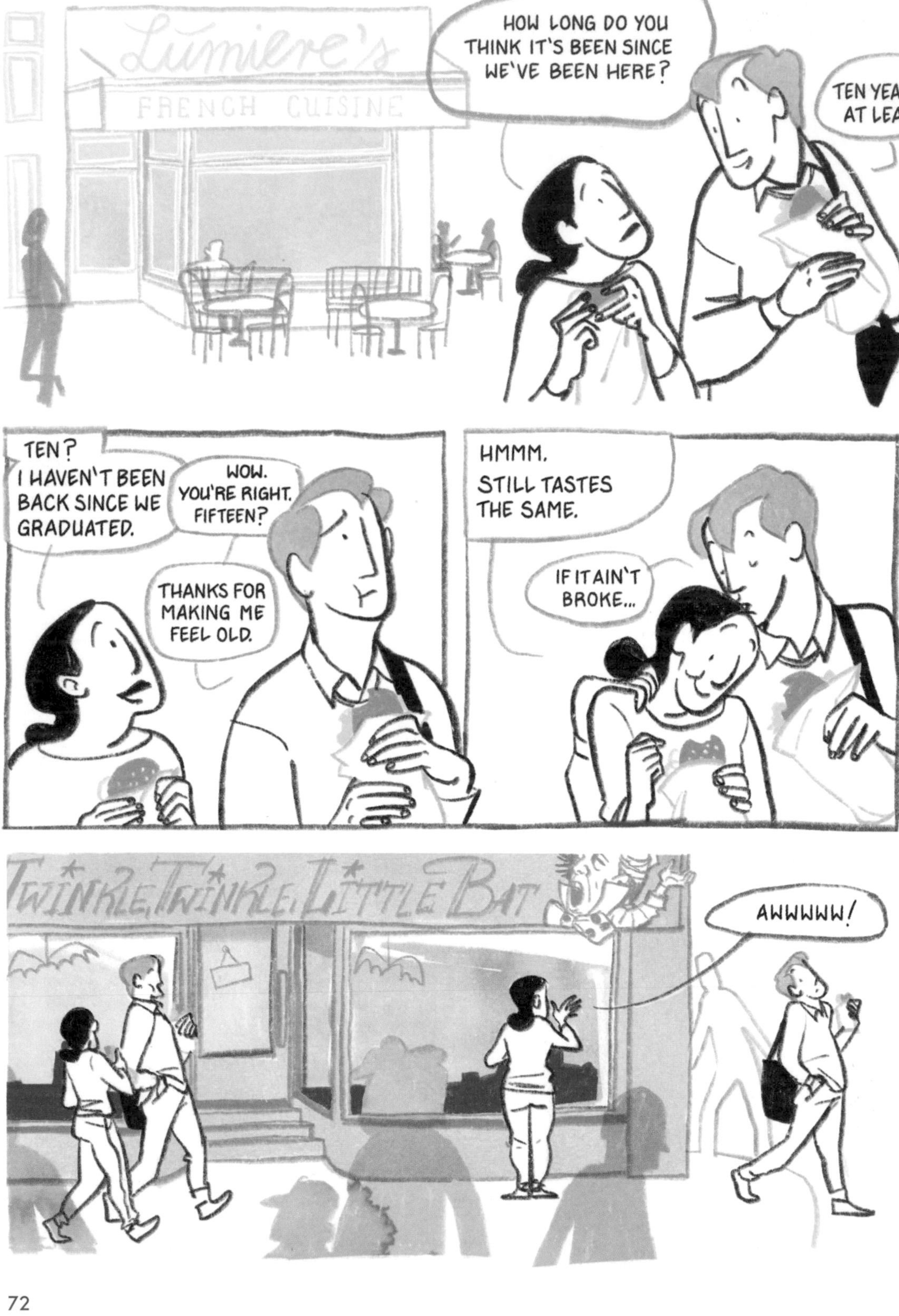
Lumiere's
FRENCH CUISINE
HOW LONG DO YOU THINK IT'S BEEN SINCE WE'VE BEEN HERE?
TEN YEARS, AT LEAST.
TEN? I HAVEN'T BEEN BACK SINCE WE GRADUATED.
WOW. YOU'RE RIGHT. FIFTEEN?
THANKS FOR MAKING ME FEEL OLD.
HMMM. STILL TASTES THE SAME.
IF IT AIN'T BROKE...
TWINKLE, TWINKLE, LITTLE BAT
AWWWWW!

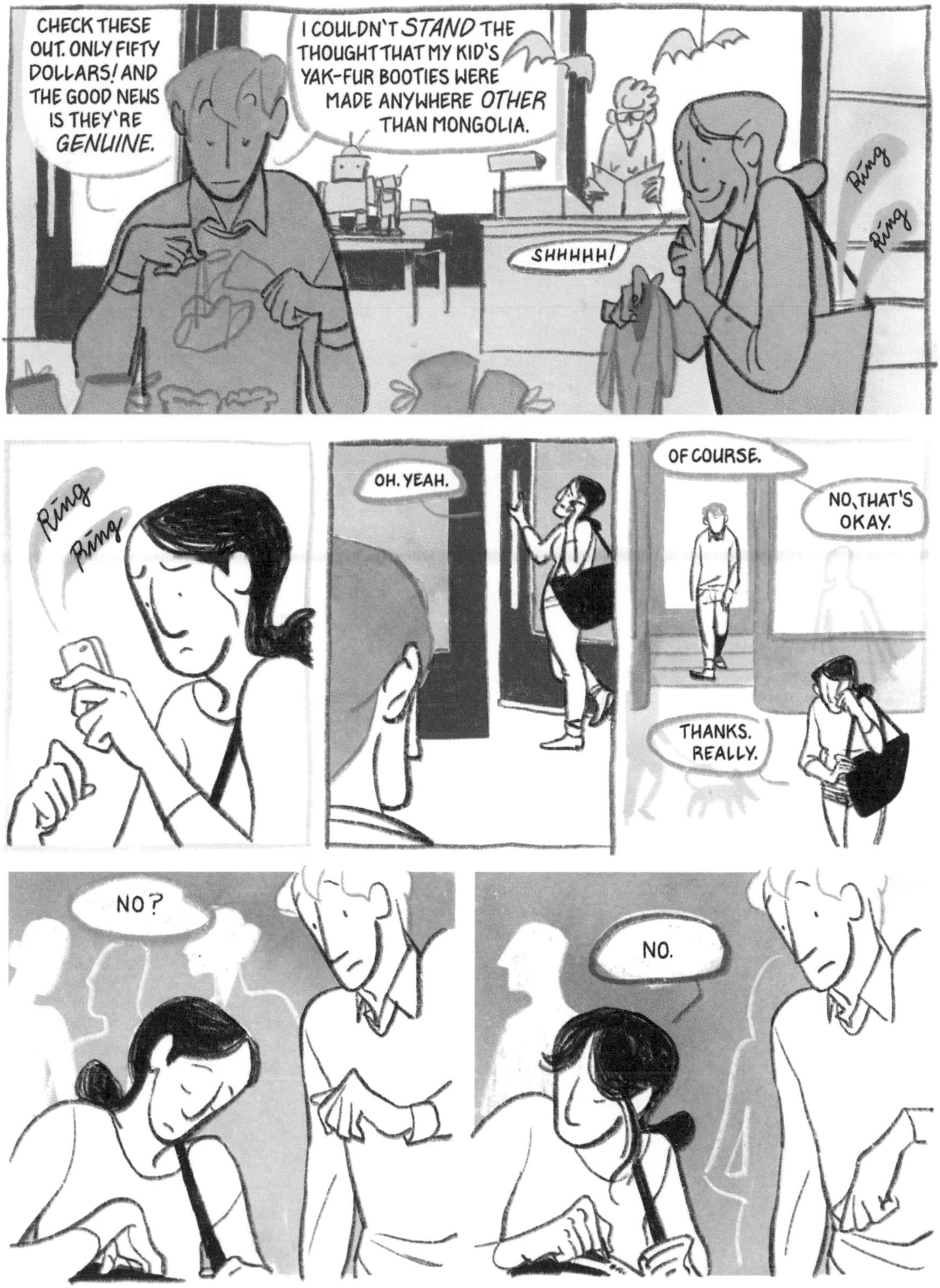
CHECK THESE OUT. ONLY FIFTY DOLLARS! AND THE GOOD NEWS IS THEY'RE GENUINE.
I COULDN'T STAND THE THOUGHT THAT MY KID'S YAK-FUR BOOTIES WERE MADE ANYWHERE OTHER THAN MONGOLIA.
SHHHHH!
Ring
Ring
Ring
Ring
OH. YEAH.
OF COURSE.
NO, THAT'S OKAY.
THANKS. REALLY.
NO?
NO.

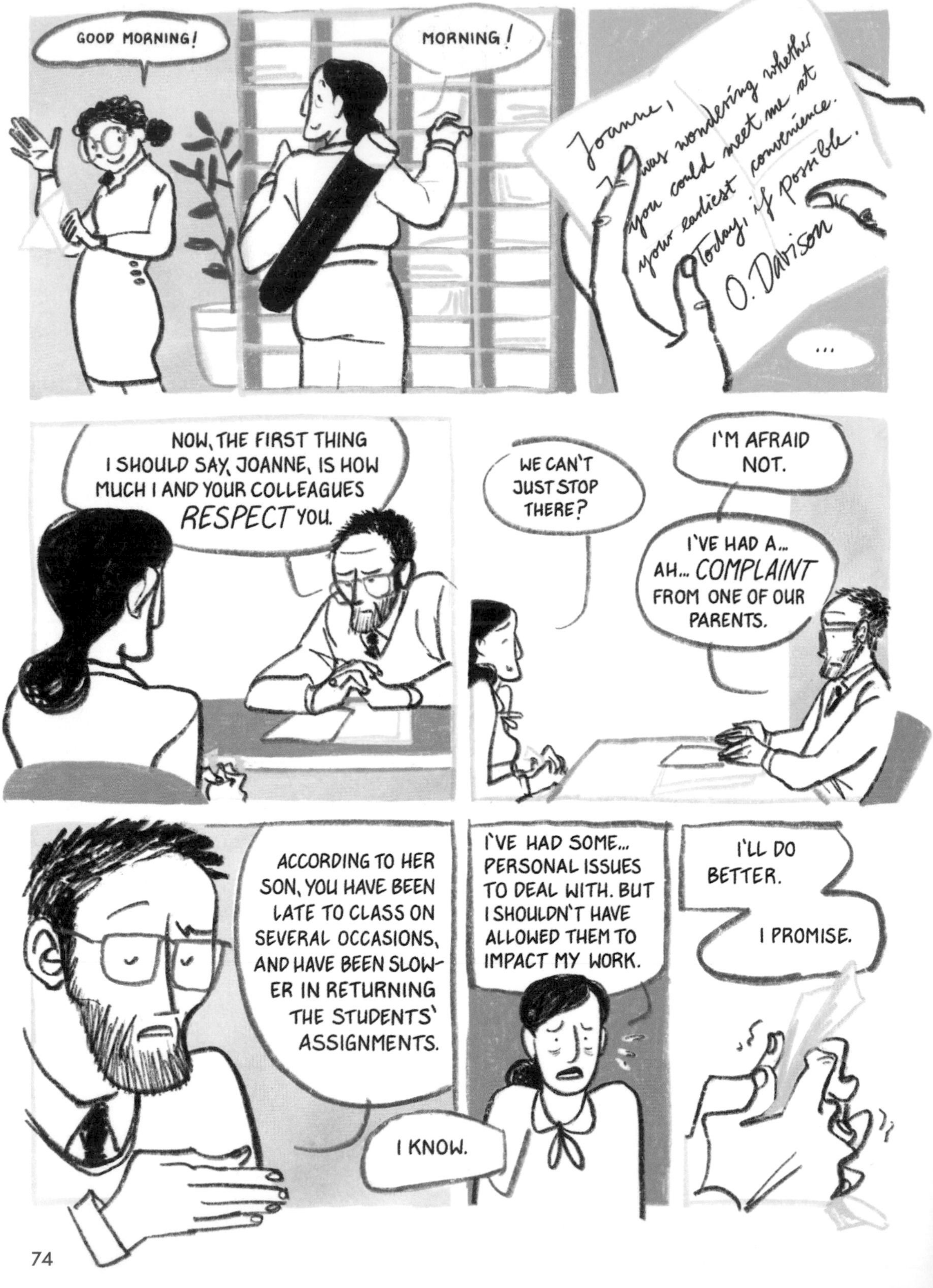
GOOD MORNING!
MORNING!
Joanne, I was wondering whether you could meet me at your earliest convenience. Today, if possible. O. Davison
...
NOW, THE FIRST THING I SHOULD SAY, JOANNE, IS HOW MUCH I AND YOUR COLLEAGUES RESPECT YOU.
WE CAN'T JUST STOP THERE?
I'M AFRAID NOT.
I'VE HAD A... AH... COMPLAINT FROM ONE OF OUR PARENTS.
ACCORDING TO HER SON, YOU HAVE BEEN LATE TO CLASS ON SEVERAL OCCASIONS, AND HAVE BEEN SLOWER IN RETURNING THE STUDENTS' ASSIGNMENTS.
I KNOW.
I'VE HAD SOME... PERSONAL ISSUES TO DEAL WITH. BUT I SHOULDN'T HAVE ALLOWED THEM TO IMPACT MY WORK.
I'LL DO BETTER.
I PROMISE.

I JUST...

I DON'T KNOW HOW--

MS HART?

MS HART?

HOW CAN I HELP?

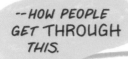
--HOW PEOPLE GET THROUGH THIS.

THEY DO IT *TOGETHER*, JO.

SORRY, MATE. COULDN'T HELP OVERHEARING.

WANNA GRAB A COFFEE?

WHAT IS SHE UP TO? HORMONES? TWO-WEEK WAIT?

SHIT.

I'M SORRY TO HEAR THAT. HOW MANY ROUNDS SO FAR?

WE JUST FOUND OUT IT DIDN'T *TAKE*.

THIS WAS OUR FIRST.

IT'S NEVER EASY, NO MATTER HOW MANY TIMES YOU GO THROUGH IT. MY WIFE AND I TRIED FOR TWO YEARS, THEN DECIDED TO GIVE UP.

WE WENT TO FIJI TO RELAX, HAD A FEW COCKTAILS. ONE THING LED TO ANOTHER. BY THE TIME WE GOT BACK, SHE WAS PREGNANT.

THANK GOD FOR BANANA DAIQUIRIS.

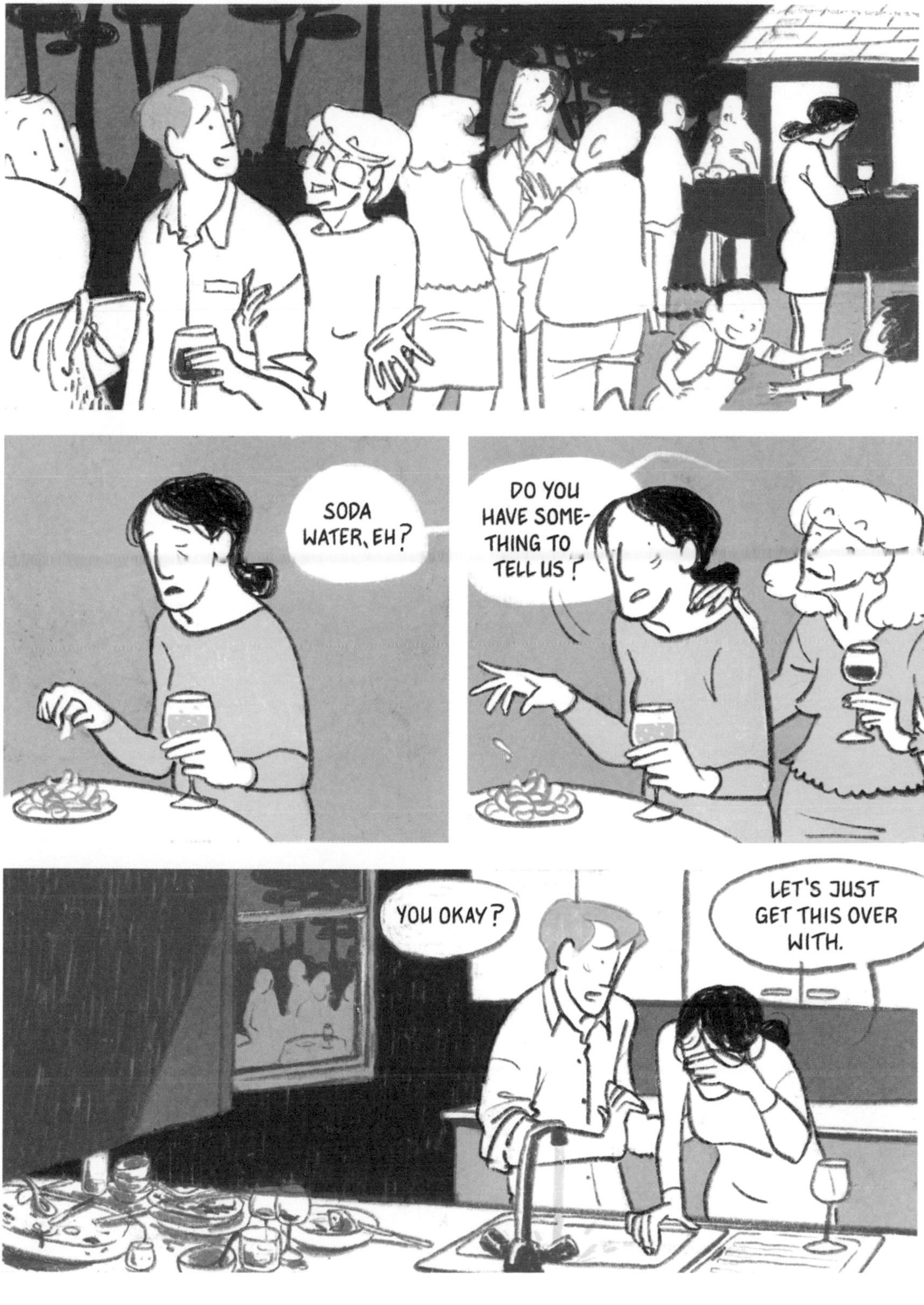
SODA WATER, EH?
DO YOU HAVE SOMETHING TO TELL US?
YOU OKAY?
LET'S JUST GET THIS OVER WITH.

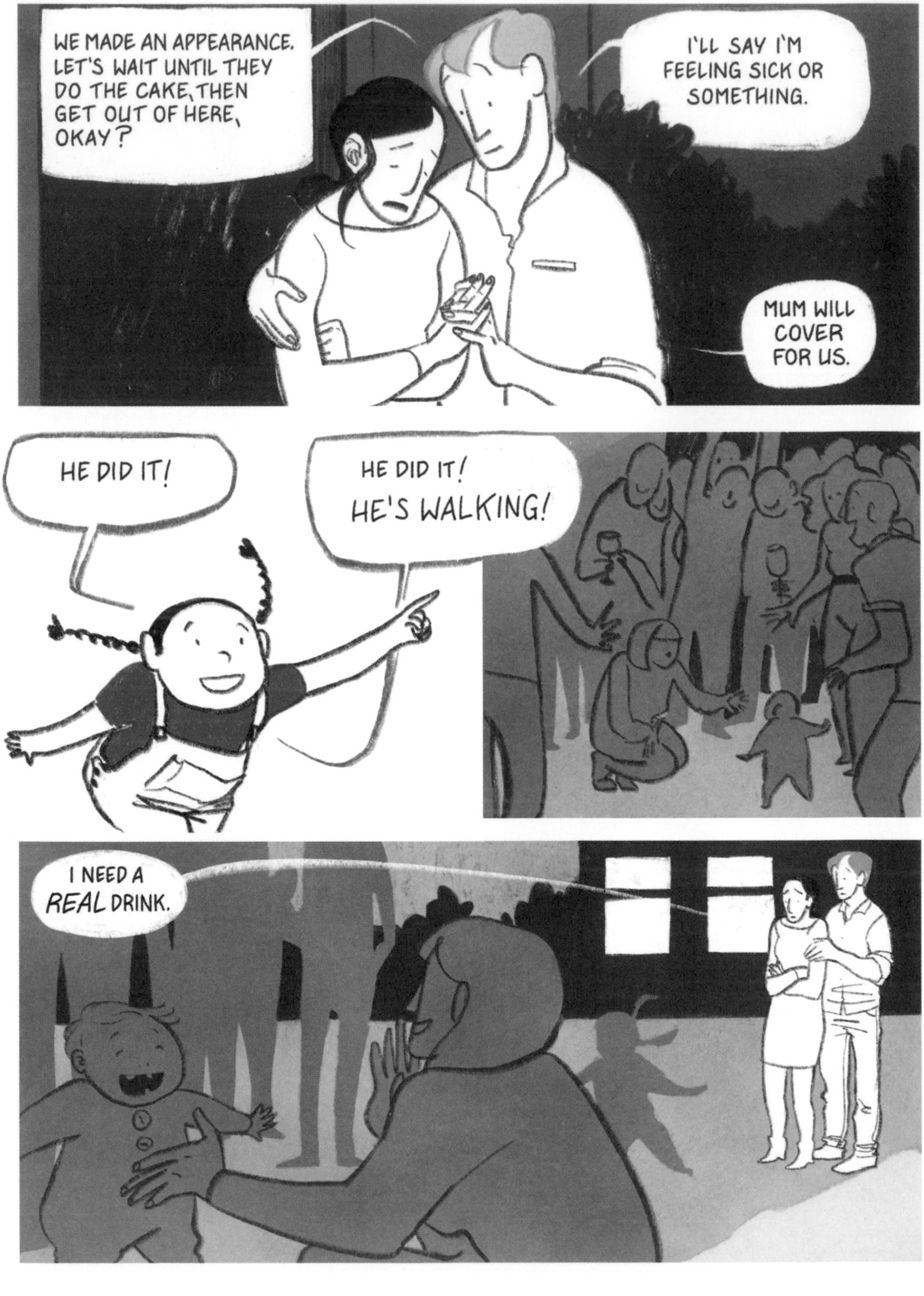
WE MADE AN APPEARANCE. LET'S WAIT UNTIL THEY DO THE CAKE, THEN GET OUT OF HERE, OKAY?
I'LL SAY I'M FEELING SICK OR SOMETHING.
MUM WILL COVER FOR US.
HE DID IT!
HE DID IT! HE'S WALKING!
I NEED A REAL DRINK.

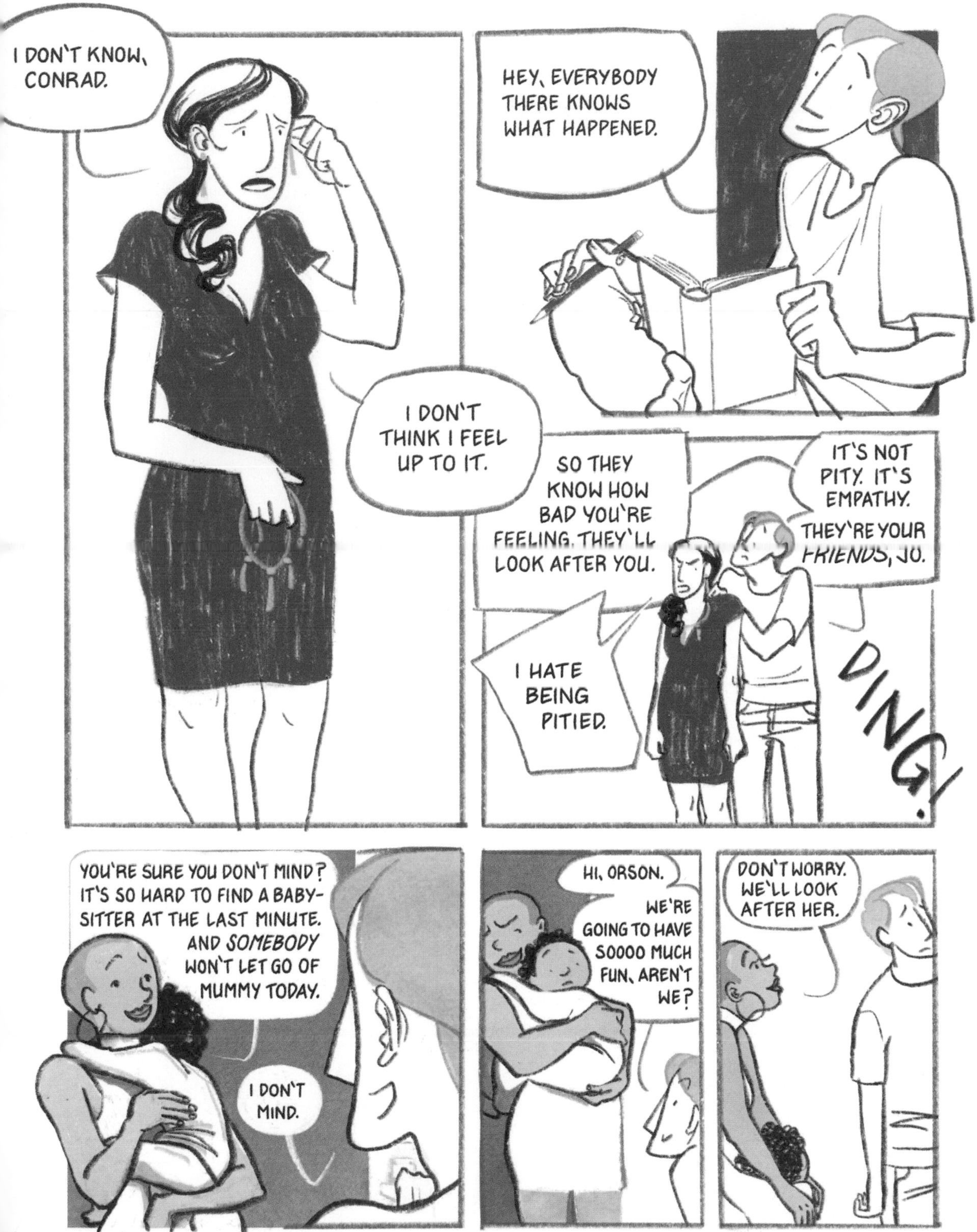
I DON'T KNOW, CONRAD.
HEY, EVERYBODY THERE KNOWS WHAT HAPPENED.
I DON'T THINK I FEEL UP TO IT.
SO THEY KNOW HOW BAD YOU'RE FEELING. THEY'LL LOOK AFTER YOU.
I HATE BEING PITIED.
IT'S NOT PITY. IT'S EMPATHY.
THEY'RE YOUR *FRIENDS*, JO.
DING!
YOU'RE SURE YOU DON'T MIND? IT'S SO HARD TO FIND A BABY-SITTER AT THE LAST MINUTE.
AND *SOMEBODY* WON'T LET GO OF MUMMY TODAY.
I DON'T MIND.
HI, ORSON.
WE'RE GOING TO HAVE SOOOO MUCH FUN, AREN'T WE?
DON'T WORRY. WE'LL LOOK AFTER HER.

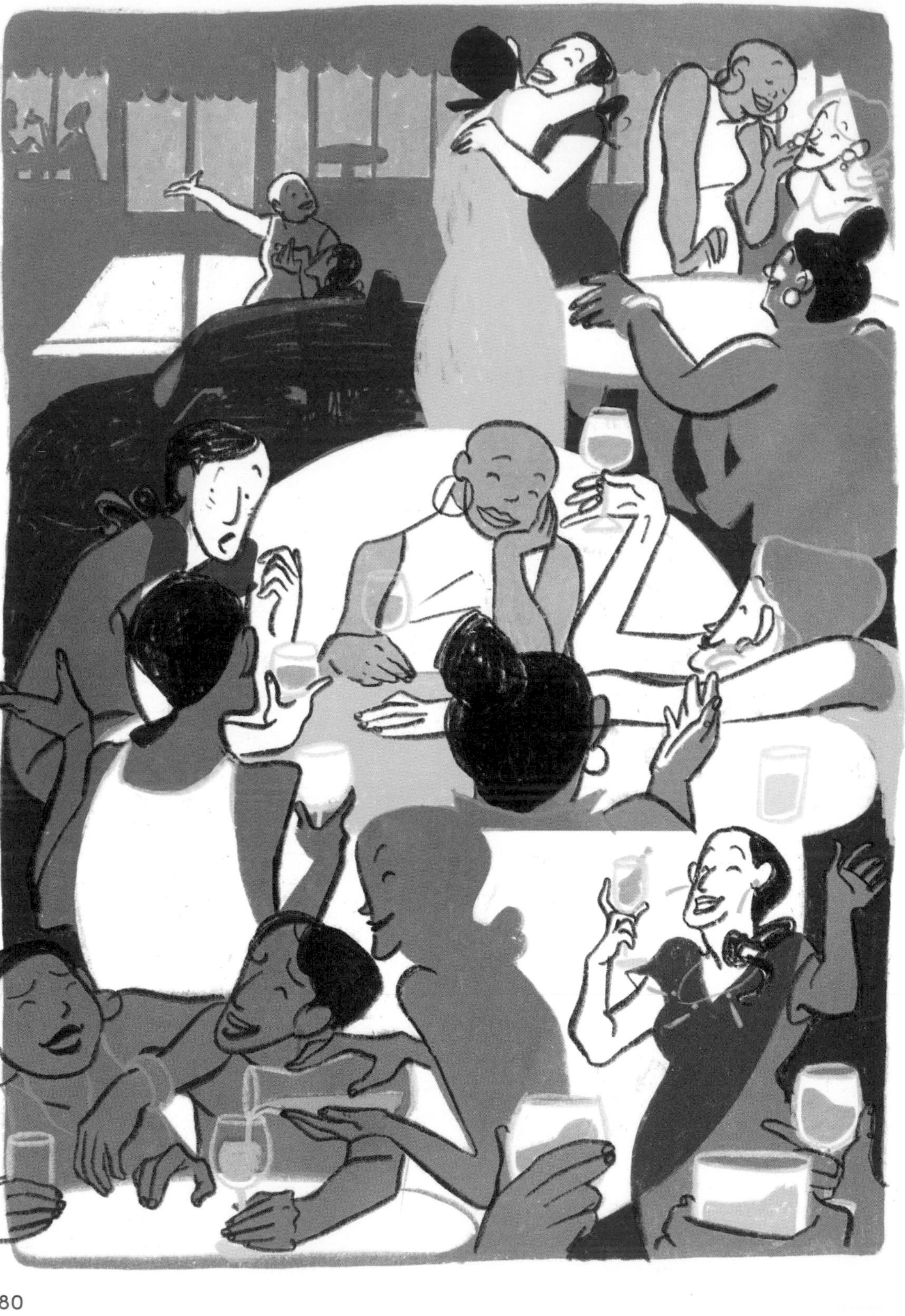

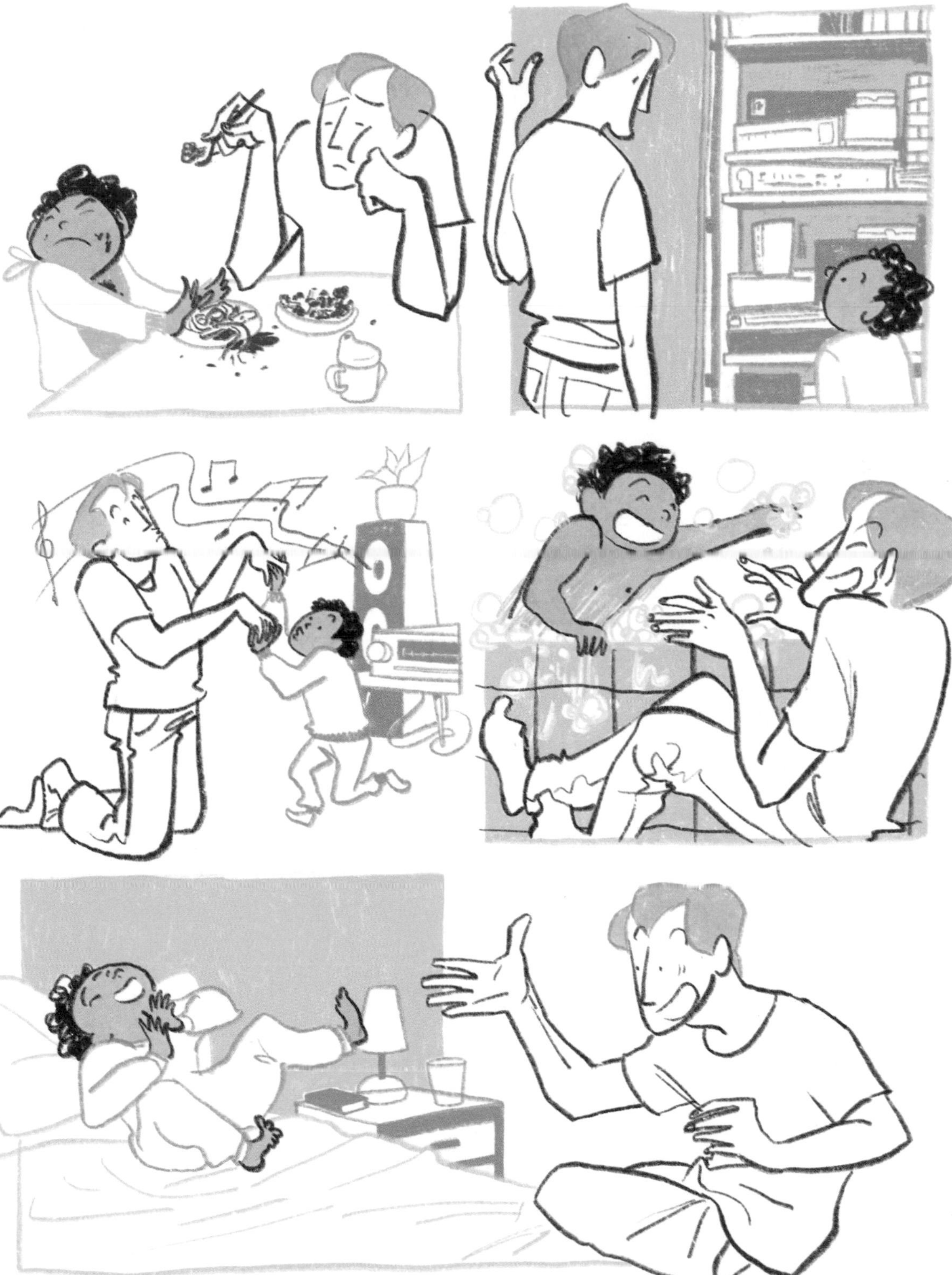

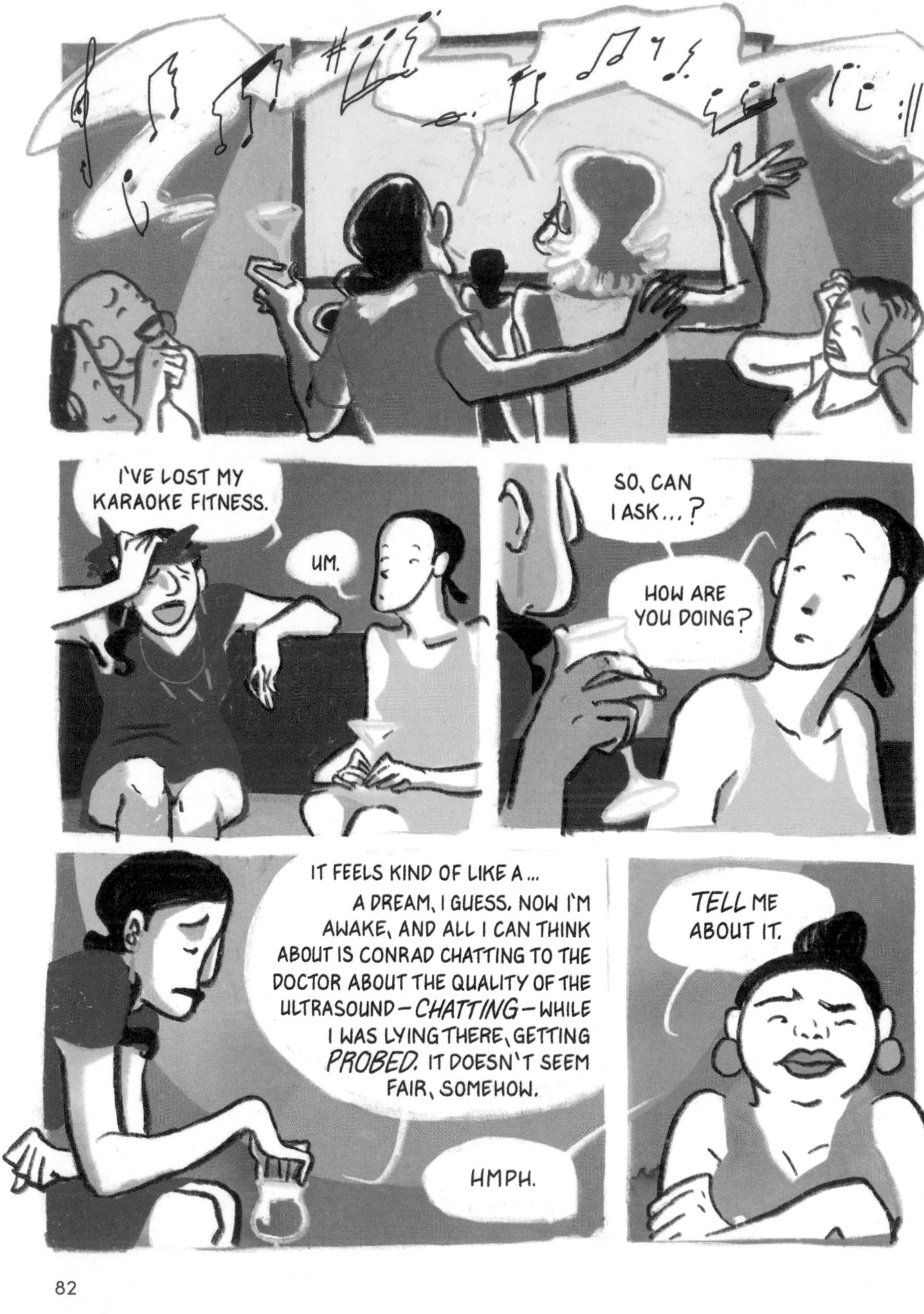
I'VE LOST MY KARAOKE FITNESS.
UM.
SO, CAN I ASK...?
HOW ARE YOU DOING?
IT FEELS KIND OF LIKE A...
A DREAM, I GUESS. NOW I'M AWAKE, AND ALL I CAN THINK ABOUT IS CONRAD CHATTING TO THE DOCTOR ABOUT THE QUALITY OF THE ULTRASOUND—CHATTING—WHILE I WAS LYING THERE, GETTING PROBED. IT DOESN'T SEEM FAIR, SOMEHOW.
HMPH.
TELL ME ABOUT IT.

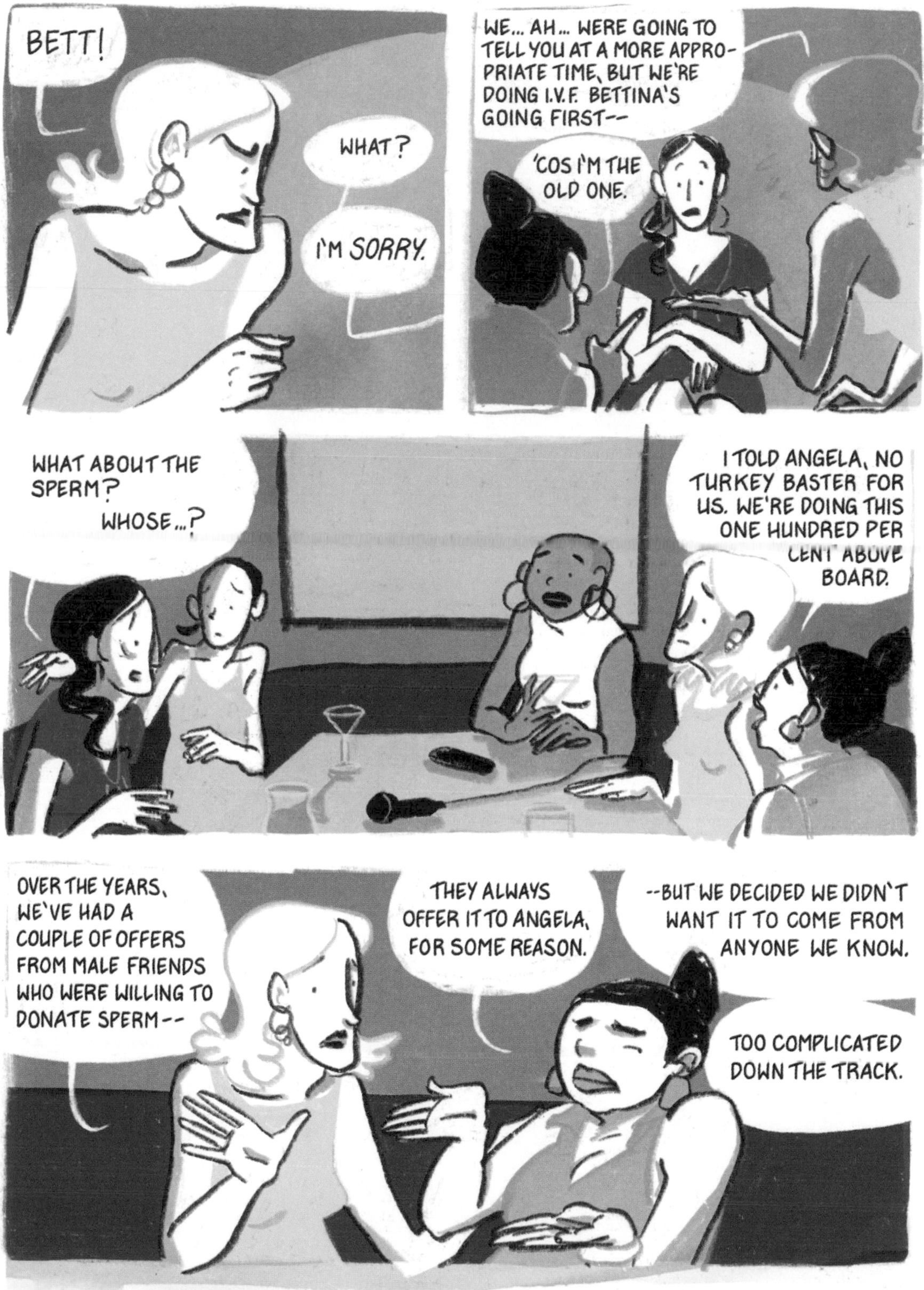
BETT!
WHAT?
I'M SORRY.
WE... AH... WERE GOING TO TELL YOU AT A MORE APPROPRIATE TIME, BUT WE'RE DOING I.V.F. BETTINA'S GOING FIRST--
'COS I'M THE OLD ONE.
WHAT ABOUT THE SPERM? WHOSE...?
I TOLD ANGELA, NO TURKEY BASTER FOR US. WE'RE DOING THIS ONE HUNDRED PER CENT ABOVE BOARD.
OVER THE YEARS, WE'VE HAD A COUPLE OF OFFERS FROM MALE FRIENDS WHO WERE WILLING TO DONATE SPERM--
THEY ALWAYS OFFER IT TO ANGELA, FOR SOME REASON.
--BUT WE DECIDED WE DIDN'T WANT IT TO COME FROM ANYONE WE KNOW.
TOO COMPLICATED DOWN THE TRACK.

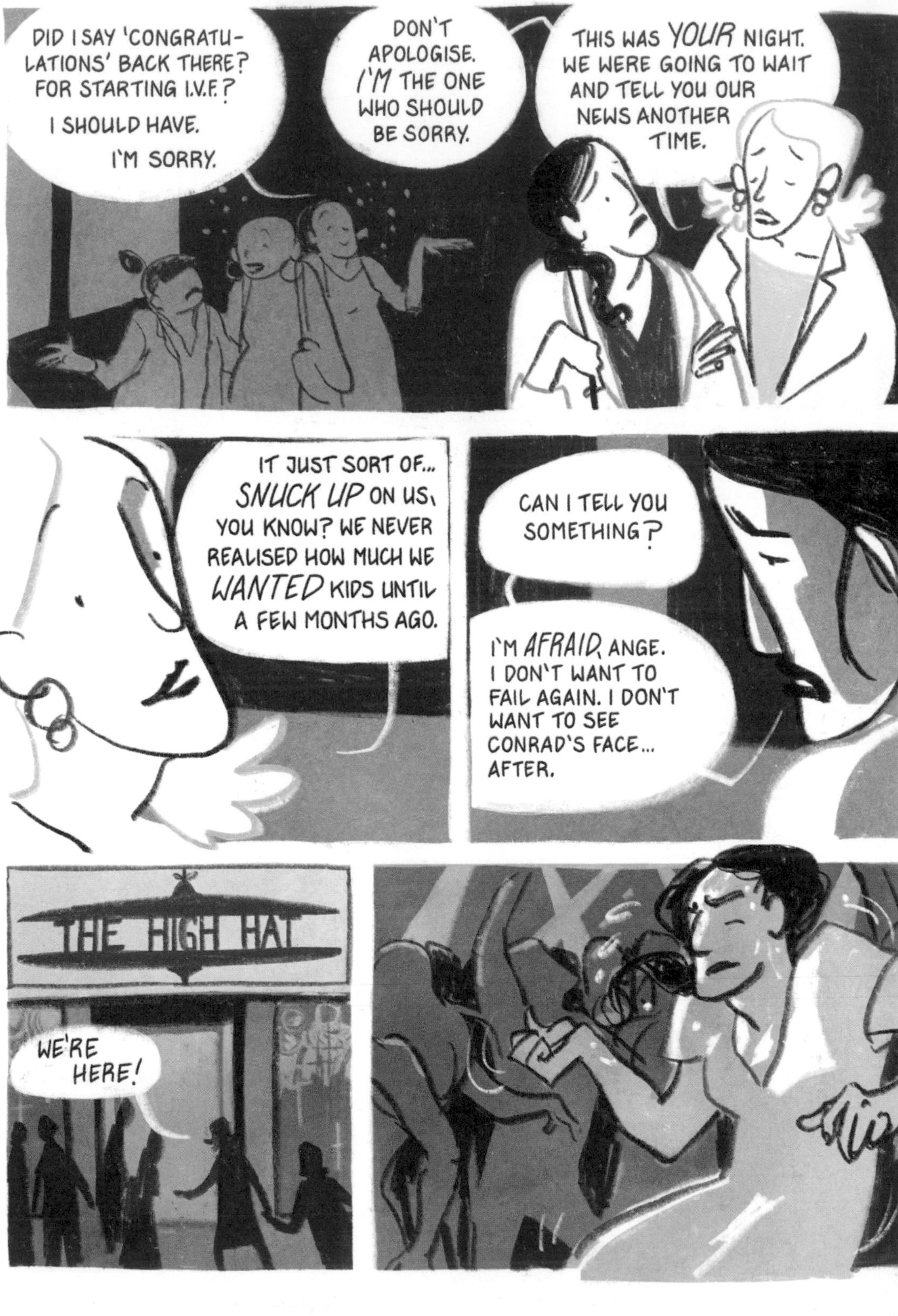
DID I SAY 'CONGRATULATIONS' BACK THERE? FOR STARTING I.V.F.? I SHOULD HAVE. I'M SORRY.
DON'T APOLOGISE. I'M THE ONE WHO SHOULD BE SORRY.
THIS WAS YOUR NIGHT. WE WERE GOING TO WAIT AND TELL YOU OUR NEWS ANOTHER TIME.
IT JUST SORT OF... SNUCK UP ON US, YOU KNOW? WE NEVER REALISED HOW MUCH WE WANTED KIDS UNTIL A FEW MONTHS AGO.
CAN I TELL YOU SOMETHING?
I'M AFRAID, ANGE. I DON'T WANT TO FAIL AGAIN. I DON'T WANT TO SEE CONRAD'S FACE... AFTER.
THE HIGH HAT
WE'RE HERE!

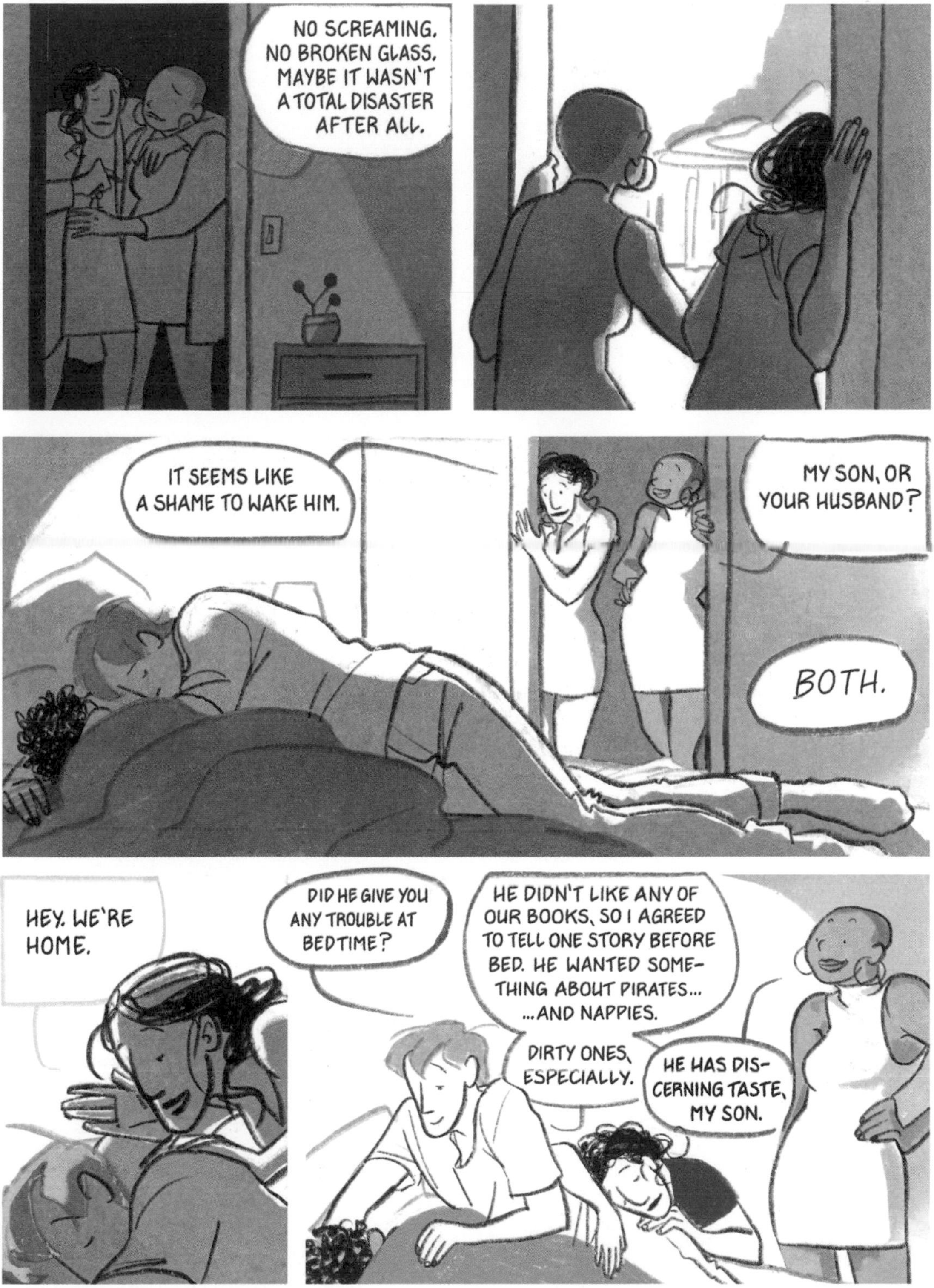
NO SCREAMING. NO BROKEN GLASS. MAYBE IT WASN'T A TOTAL DISASTER AFTER ALL.
IT SEEMS LIKE A SHAME TO WAKE HIM.
MY SON, OR YOUR HUSBAND?
BOTH.
HEY. WE'RE HOME.
DID HE GIVE YOU ANY TROUBLE AT BEDTIME?
HE DIDN'T LIKE ANY OF OUR BOOKS, SO I AGREED TO TELL ONE STORY BEFORE BED. HE WANTED SOMETHING ABOUT PIRATES... ...AND NAPPIES.
DIRTY ONES, ESPECIALLY.
HE HAS DISCERNING TASTE, MY SON.

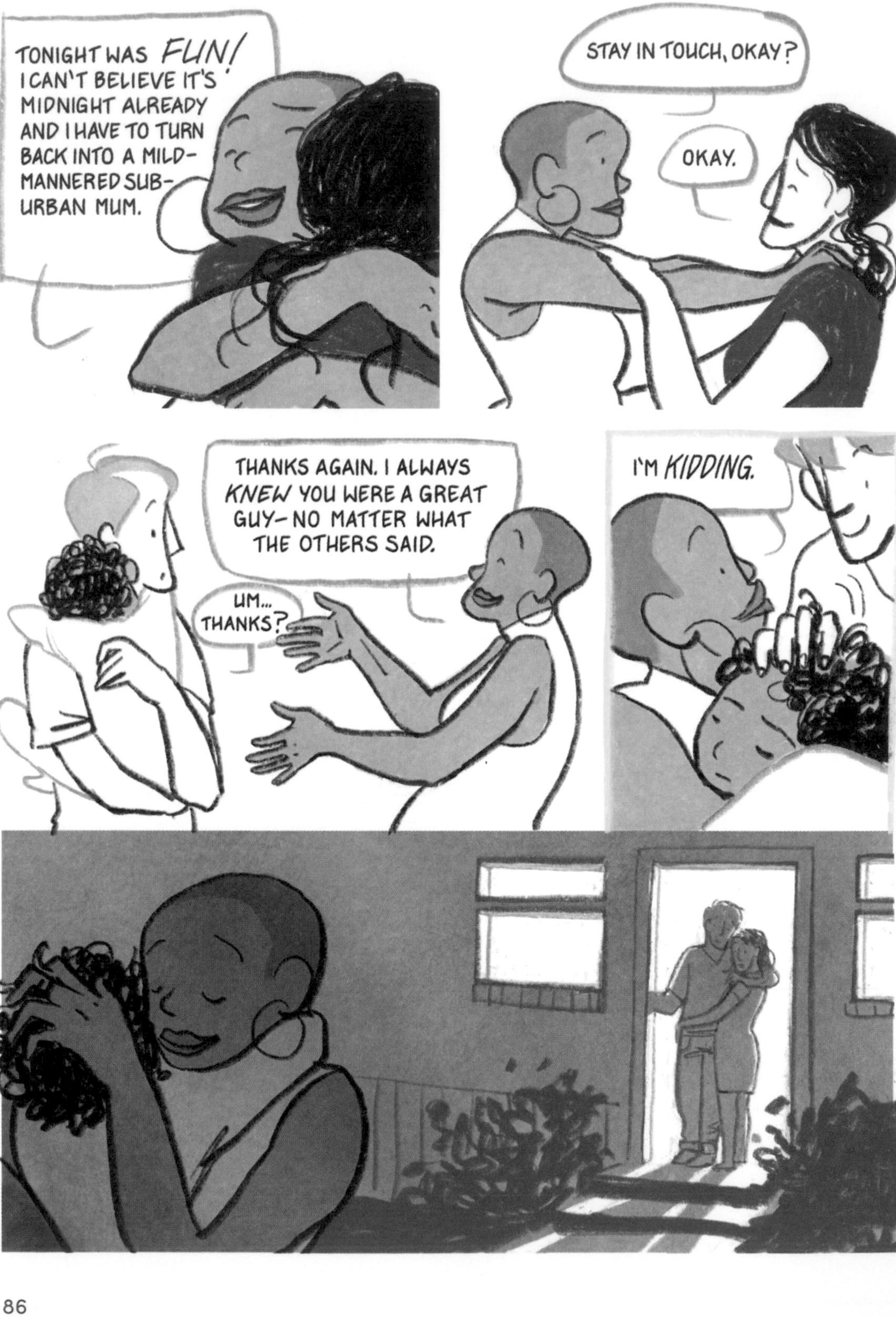
TONIGHT WAS FUN! I CAN'T BELIEVE IT'S MIDNIGHT ALREADY AND I HAVE TO TURN BACK INTO A MILD-MANNERED SUB-URBAN MUM.
STAY IN TOUCH, OKAY?
OKAY.
THANKS AGAIN. I ALWAYS KNEW YOU WERE A GREAT GUY– NO MATTER WHAT THE OTHERS SAID.
UM... THANKS?
I'M KIDDING.

CHAPTER 5:

MERRY-
GO-
ROUND

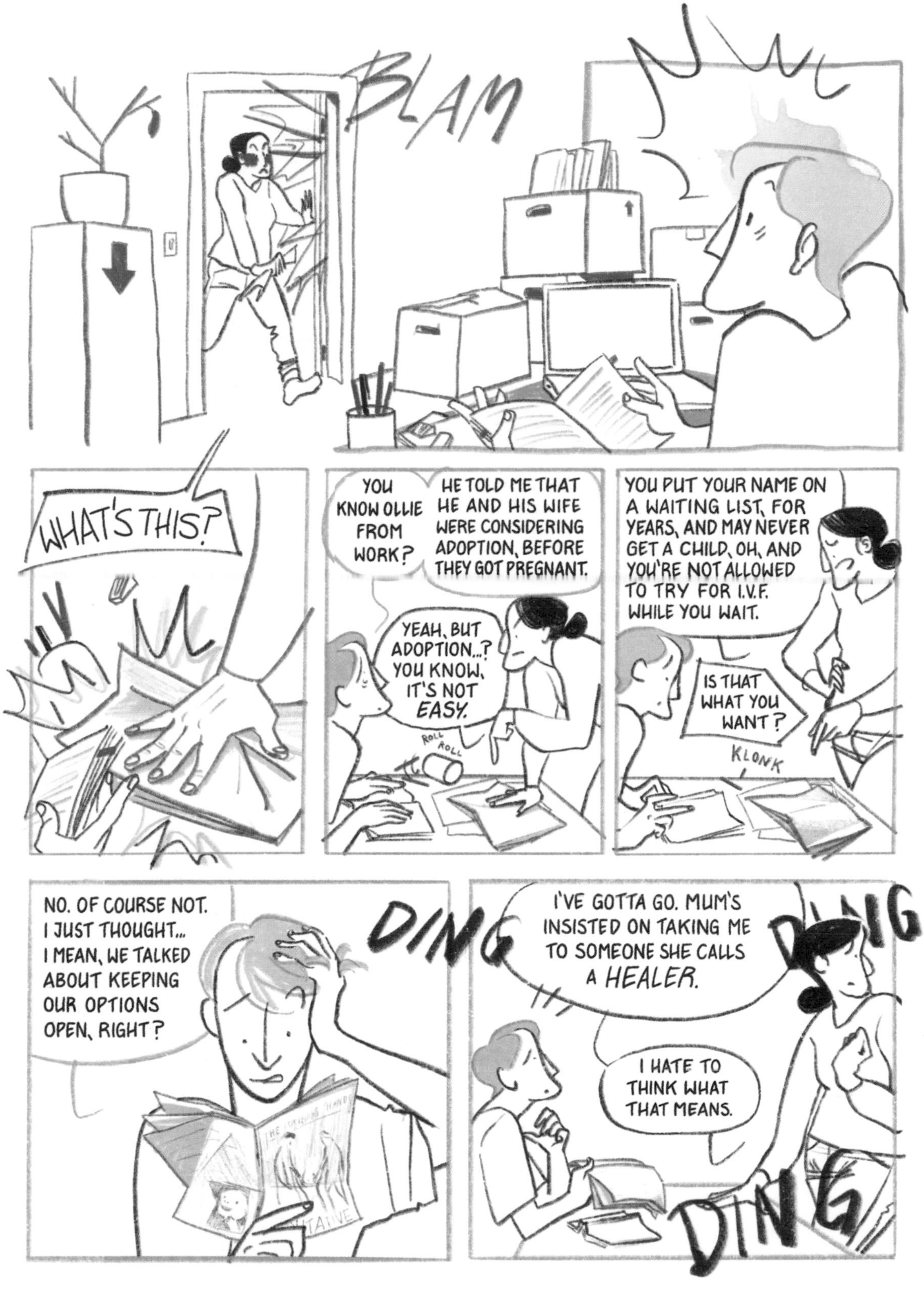
BLAM
WHAT'S THIS?
YOU KNOW OLLIE FROM WORK?
HE TOLD ME THAT HE AND HIS WIFE WERE CONSIDERING ADOPTION, BEFORE THEY GOT PREGNANT.
YEAH, BUT ADOPTION...? YOU KNOW, IT'S NOT EASY.
ROLL ROLL
YOU PUT YOUR NAME ON A WAITING LIST, FOR YEARS, AND MAY NEVER GET A CHILD, OH, AND YOU'RE NOT ALLOWED TO TRY FOR I.V.F. WHILE YOU WAIT.
IS THAT WHAT YOU WANT?
KLONK
NO. OF COURSE NOT. I JUST THOUGHT... I MEAN, WE TALKED ABOUT KEEPING OUR OPTIONS OPEN, RIGHT?
DING
I'VE GOTTA GO. MUM'S INSISTED ON TAKING ME TO SOMEONE SHE CALLS A HEALER.
DING
I HATE TO THINK WHAT THAT MEANS.
DING

SO TELL ME WHO THIS PERSON IS AGAIN. WHAT HER *CREDENTIALS* ARE, I MEAN.

SEE? THIS IS WHAT I'M *SAYING*.

YOU'RE SO INDOCTRINATED BY WESTERN MEDICINE, YOU'RE NOT EVEN WILLING TO *CONSIDER*--

ALRIGHT, ALRIGHT.

THE GIRLS AT WORK TOLD ME SHE'S THE *BEST.* A REAL *FERTILITY SPECIALIST.*

BESIDES, IT DOESN'T HURT TO *TRY* SOME ALTERNATIVES, DOES IT?

YOU MUST BE JOANNE.

IT'S SO LOVELY TO MEET YOU. WE'RE GOING TO HAVE A WONDERFUL MORNING, I *FEEL* IT ALREADY.

ON THE PHONE, YOUR MUM MENTIONED THAT YOU'RE DOING I.V.F.

HOW ARE YOU FINDING THE PROCESS?

STRANGE.

DISAPPOINTING, I GUESS.

HMMM...

WE JUST FINISHED OUR FIRST ROUND.

NO LUCK.

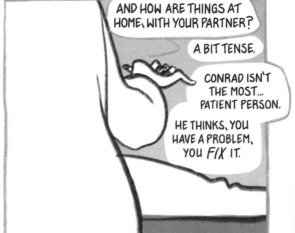

AND HOW ARE THINGS AT HOME, WITH YOUR PARTNER?

A BIT TENSE.

CONRAD ISN'T THE MOST... PATIENT PERSON.

HE THINKS, YOU HAVE A PROBLEM, YOU *FIX* IT.

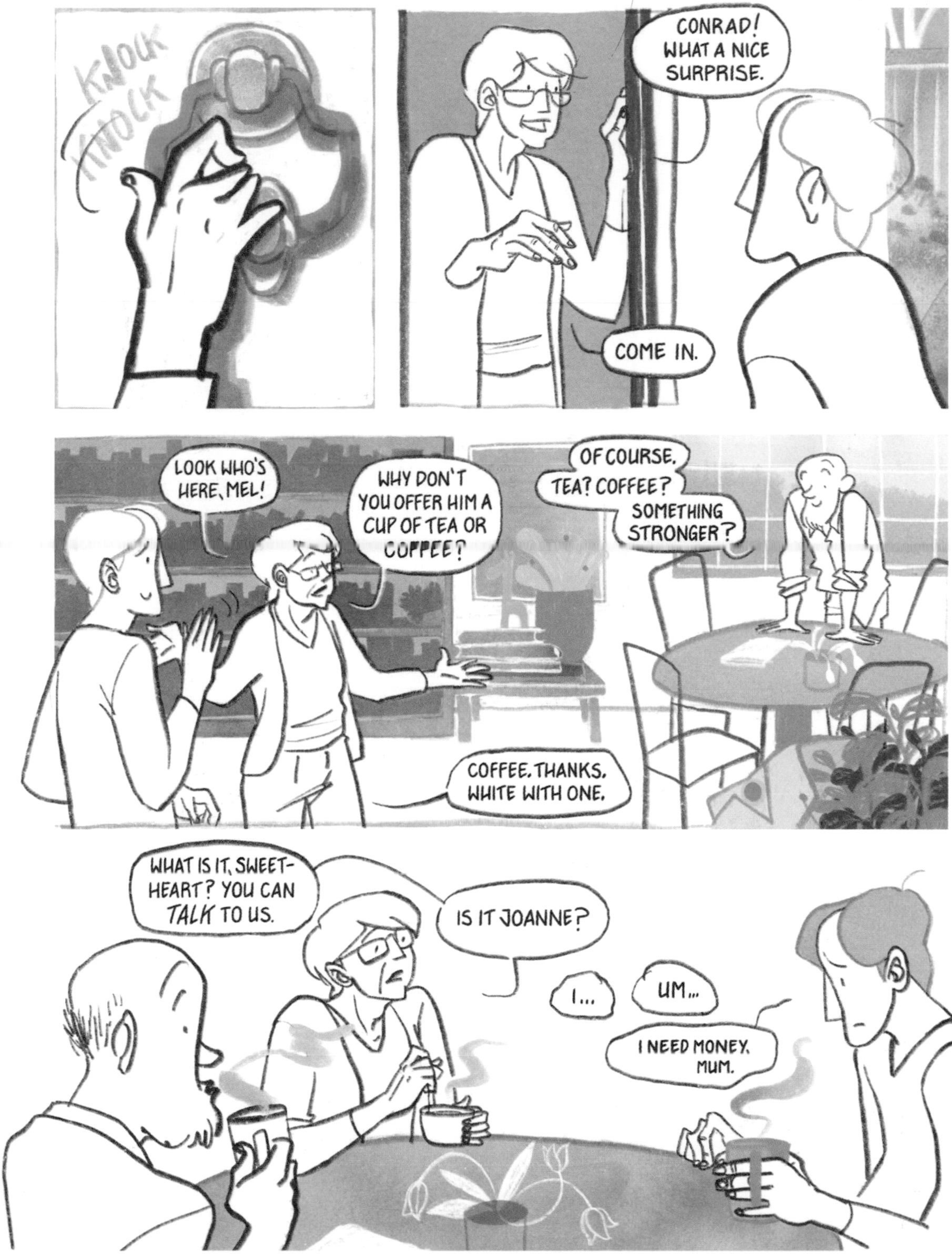
KNOCK
KNOCK
CONRAD! WHAT A NICE SURPRISE.
COME IN.
LOOK WHO'S HERE, MEL!
WHY DON'T YOU OFFER HIM A CUP OF TEA OR COFFEE?
OF COURSE. TEA? COFFEE?
SOMETHING STRONGER?
COFFEE. THANKS. WHITE WITH ONE.
WHAT IS IT, SWEET-HEART? YOU CAN *TALK* TO US.
IS IT JOANNE?
I...
UM...
I NEED MONEY. MUM.

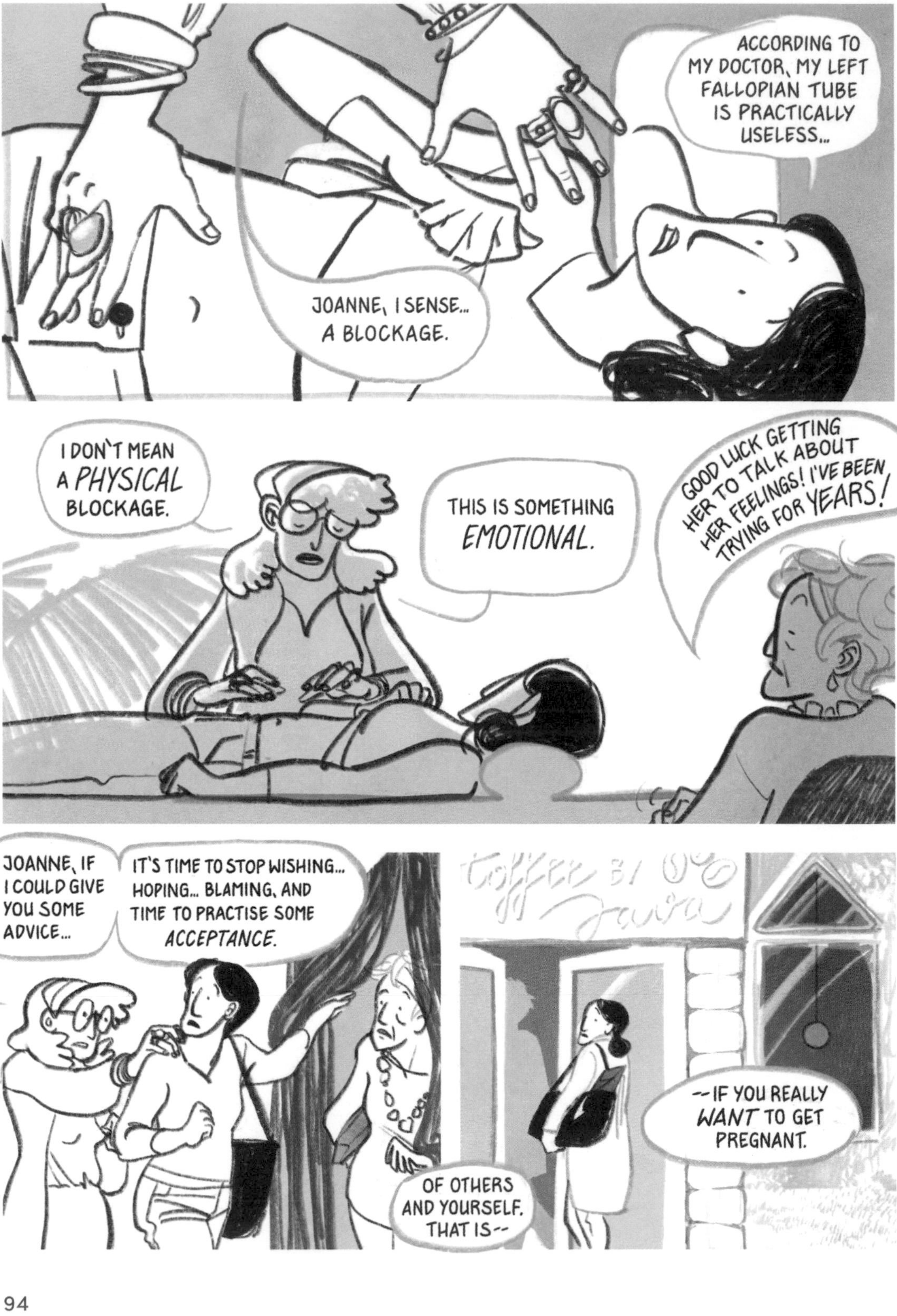
ACCORDING TO MY DOCTOR, MY LEFT FALLOPIAN TUBE IS PRACTICALLY USELESS...
JOANNE, I SENSE... A BLOCKAGE.
I DON'T MEAN A PHYSICAL BLOCKAGE.
THIS IS SOMETHING EMOTIONAL.
GOOD LUCK GETTING HER TO TALK ABOUT HER FEELINGS! I'VE BEEN TRYING FOR YEARS!
JOANNE, IF I COULD GIVE YOU SOME ADVICE...
IT'S TIME TO STOP WISHING... HOPING... BLAMING, AND TIME TO PRACTISE SOME ACCEPTANCE.
OF OTHERS AND YOURSELF. THAT IS--
--IF YOU REALLY WANT TO GET PREGNANT.

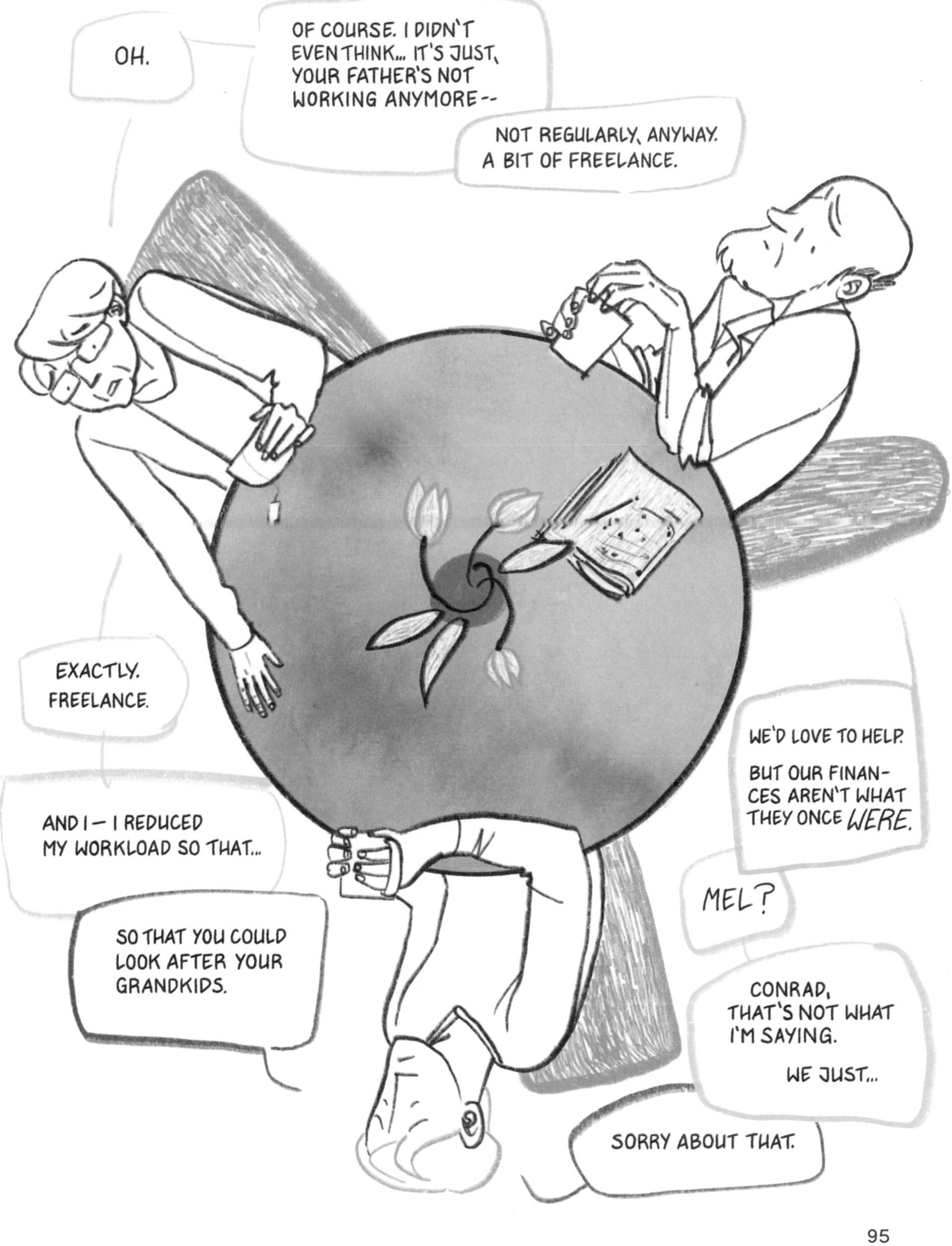
OH.
OF COURSE. I DIDN'T EVEN THINK... IT'S JUST, YOUR FATHER'S NOT WORKING ANYMORE--
NOT REGULARLY, ANYWAY. A BIT OF FREELANCE.
EXACTLY. FREELANCE.
AND I — I REDUCED MY WORKLOAD SO THAT...
SO THAT YOU COULD LOOK AFTER YOUR GRANDKIDS.
WE'D LOVE TO HELP.
BUT OUR FINANCES AREN'T WHAT THEY ONCE *WERE*.
MEL?
CONRAD, THAT'S NOT WHAT I'M SAYING.
WE JUST...
SORRY ABOUT THAT.

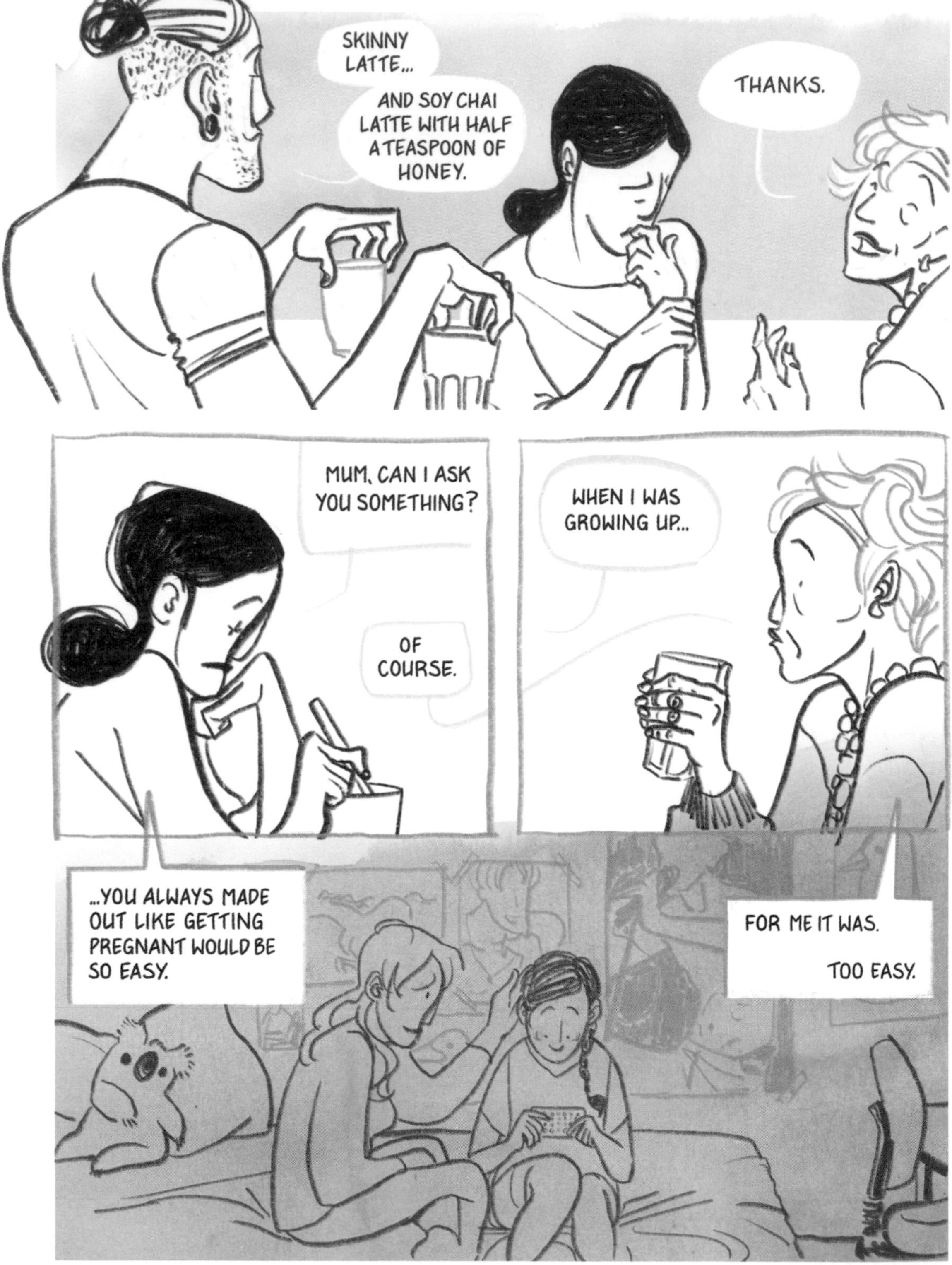
SKINNY LATTE...
AND SOY CHAI LATTE WITH HALF A TEASPOON OF HONEY.
THANKS.
MUM, CAN I ASK YOU SOMETHING?
OF COURSE.
WHEN I WAS GROWING UP...
...YOU ALWAYS MADE OUT LIKE GETTING PREGNANT WOULD BE SO EASY.
FOR ME IT WAS.
TOO EASY.

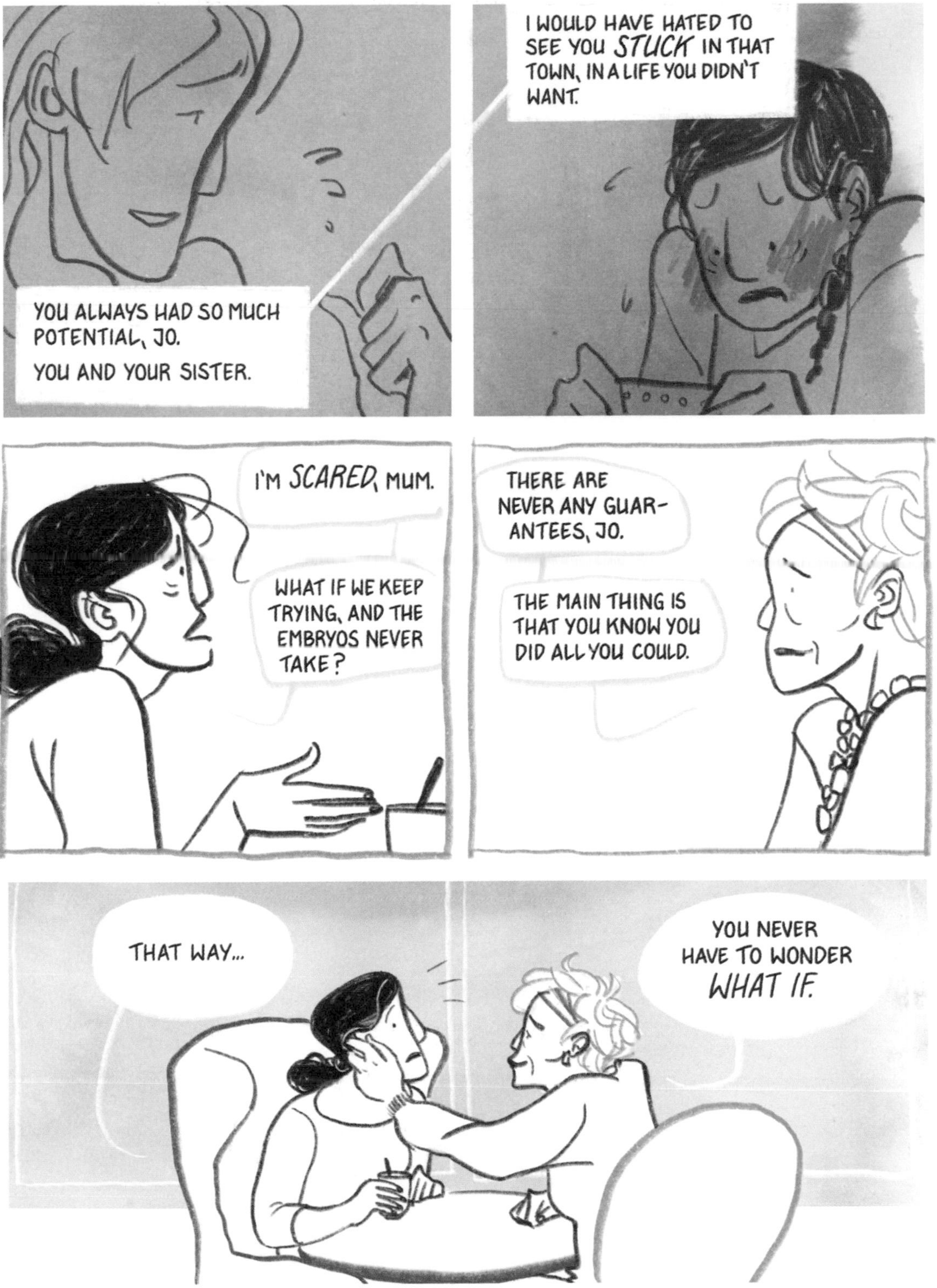
YOU ALWAYS HAD SO MUCH POTENTIAL, JO.
YOU AND YOUR SISTER.
I WOULD HAVE HATED TO SEE YOU STUCK IN THAT TOWN, IN A LIFE YOU DIDN'T WANT.
I'M SCARED, MUM.
WHAT IF WE KEEP TRYING, AND THE EMBRYOS NEVER TAKE?
THERE ARE NEVER ANY GUARANTEES, JO.
THE MAIN THING IS THAT YOU KNOW YOU DID ALL YOU COULD.
THAT WAY...
YOU NEVER HAVE TO WONDER WHAT IF.

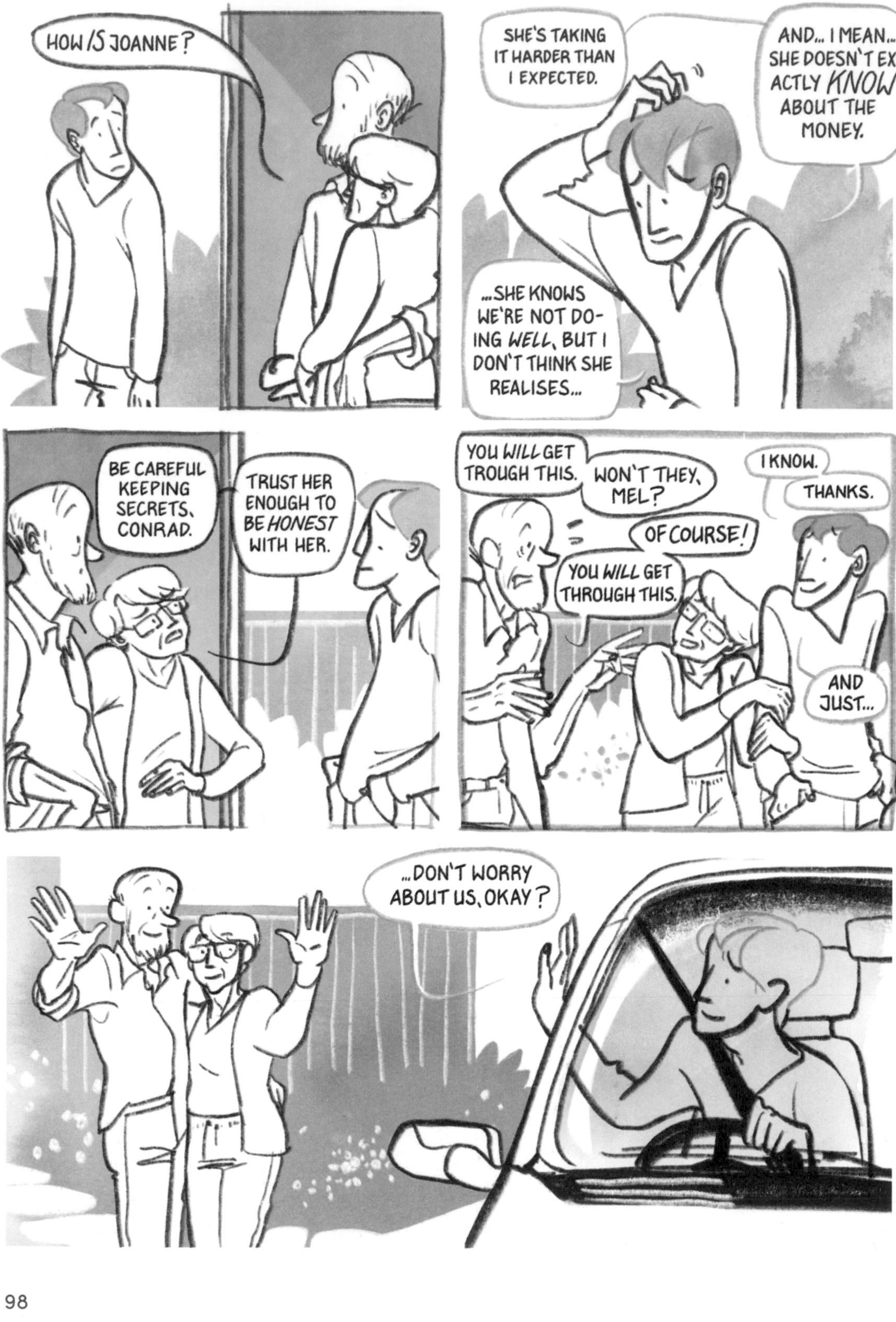
HOW IS JOANNE?
SHE'S TAKING IT HARDER THAN I EXPECTED.
AND... I MEAN... SHE DOESN'T EXACTLY KNOW ABOUT THE MONEY.
...SHE KNOWS WE'RE NOT DOING WELL, BUT I DON'T THINK SHE REALISES...
BE CAREFUL KEEPING SECRETS, CONRAD.
TRUST HER ENOUGH TO BE HONEST WITH HER.
YOU WILL GET TROUGH THIS.
WON'T THEY, MEL?
OF COURSE!
YOU WILL GET THROUGH THIS.
I KNOW.
THANKS.
AND JUST...
...DON'T WORRY ABOUT US, OKAY?

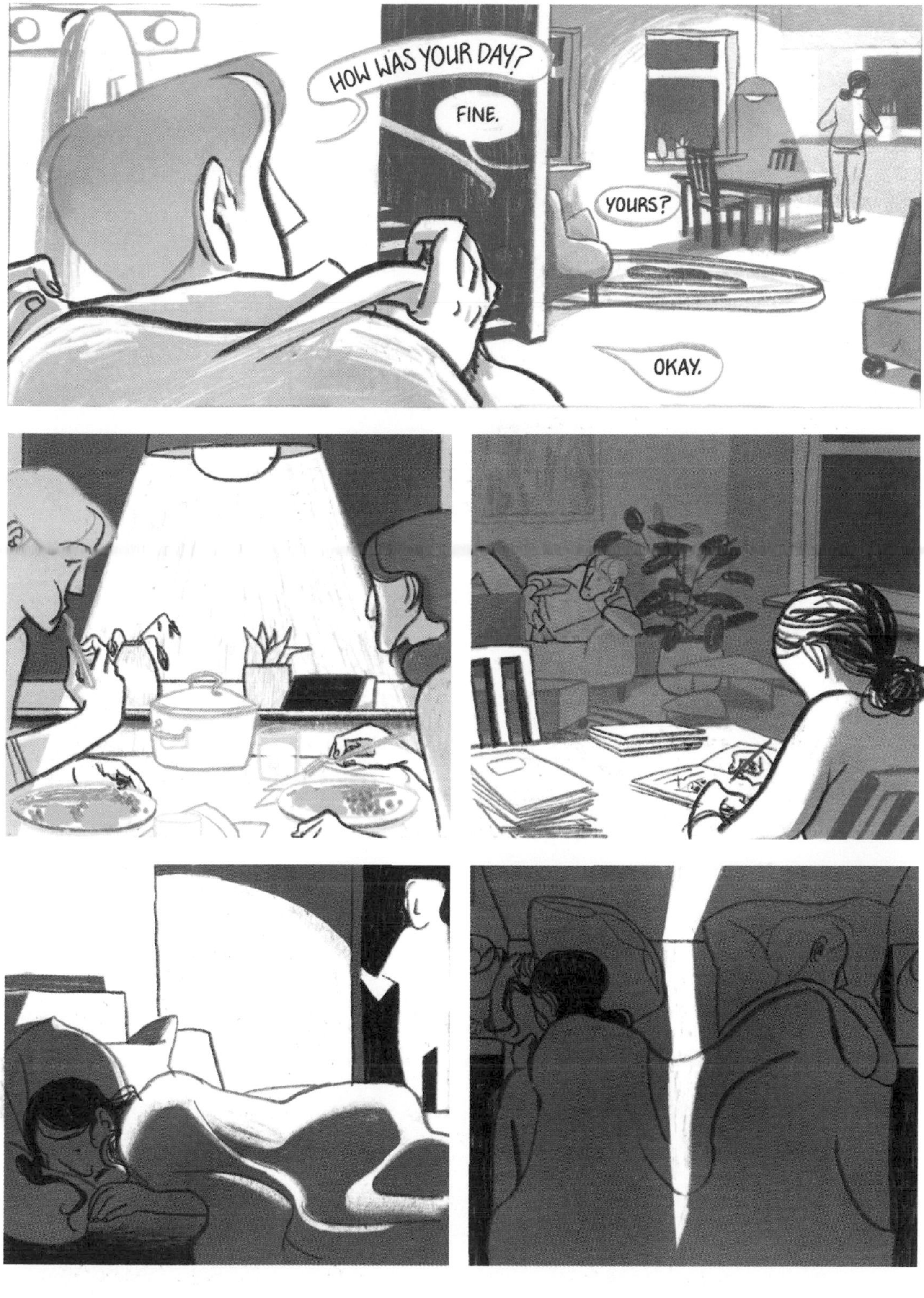
HOW WAS YOUR DAY?
FINE.
YOURS?
OKAY.

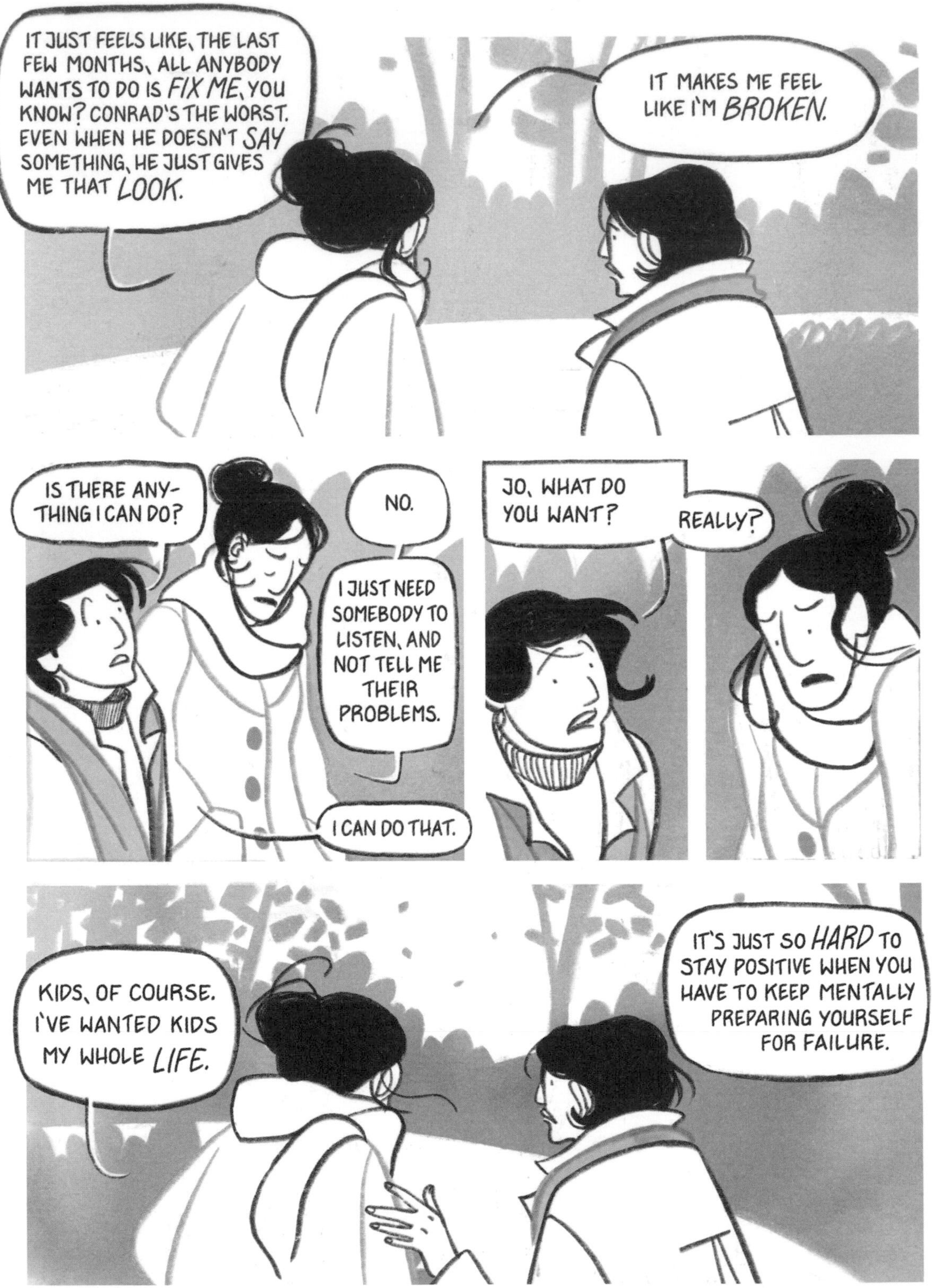
IT JUST FEELS LIKE, THE LAST FEW MONTHS, ALL ANYBODY WANTS TO DO IS FIX ME, YOU KNOW? CONRAD'S THE WORST. EVEN WHEN HE DOESN'T SAY SOMETHING, HE JUST GIVES ME THAT LOOK.
IT MAKES ME FEEL LIKE I'M BROKEN.
IS THERE ANYTHING I CAN DO?
NO.
I JUST NEED SOMEBODY TO LISTEN, AND NOT TELL ME THEIR PROBLEMS.
I CAN DO THAT.
JO, WHAT DO YOU WANT?
REALLY?
KIDS, OF COURSE. I'VE WANTED KIDS MY WHOLE LIFE.
IT'S JUST SO HARD TO STAY POSITIVE WHEN YOU HAVE TO KEEP MENTALLY PREPARING YOURSELF FOR FAILURE.

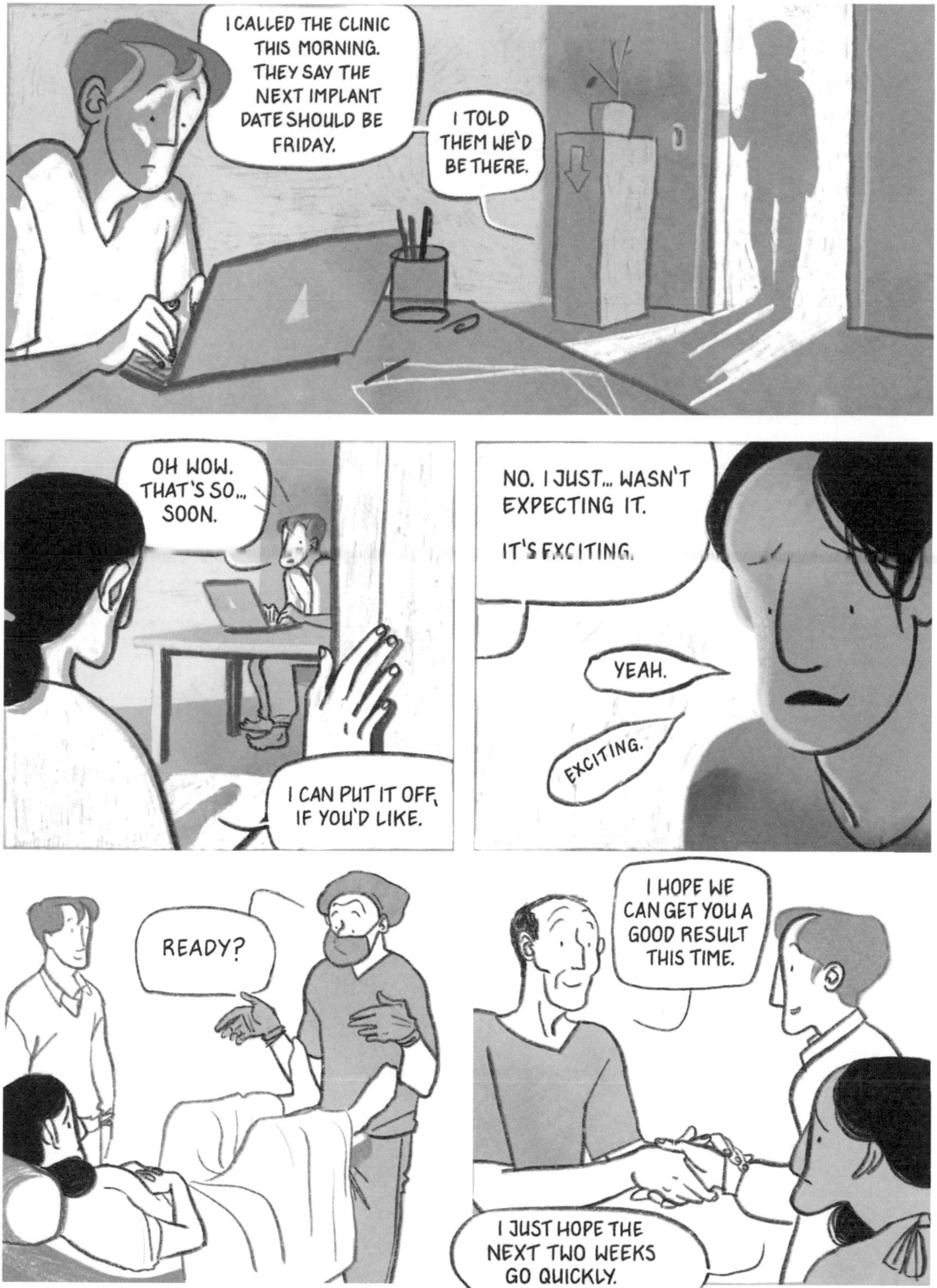
I CALLED THE CLINIC THIS MORNING. THEY SAY THE NEXT IMPLANT DATE SHOULD BE FRIDAY.
I TOLD THEM WE'D BE THERE.
OH WOW. THAT'S SO... SOON.
I CAN PUT IT OFF, IF YOU'D LIKE.
NO. I JUST... WASN'T EXPECTING IT.
IT'S EXCITING.
YEAH.
EXCITING.
READY?
I HOPE WE CAN GET YOU A GOOD RESULT THIS TIME.
I JUST HOPE THE NEXT TWO WEEKS GO QUICKLY.

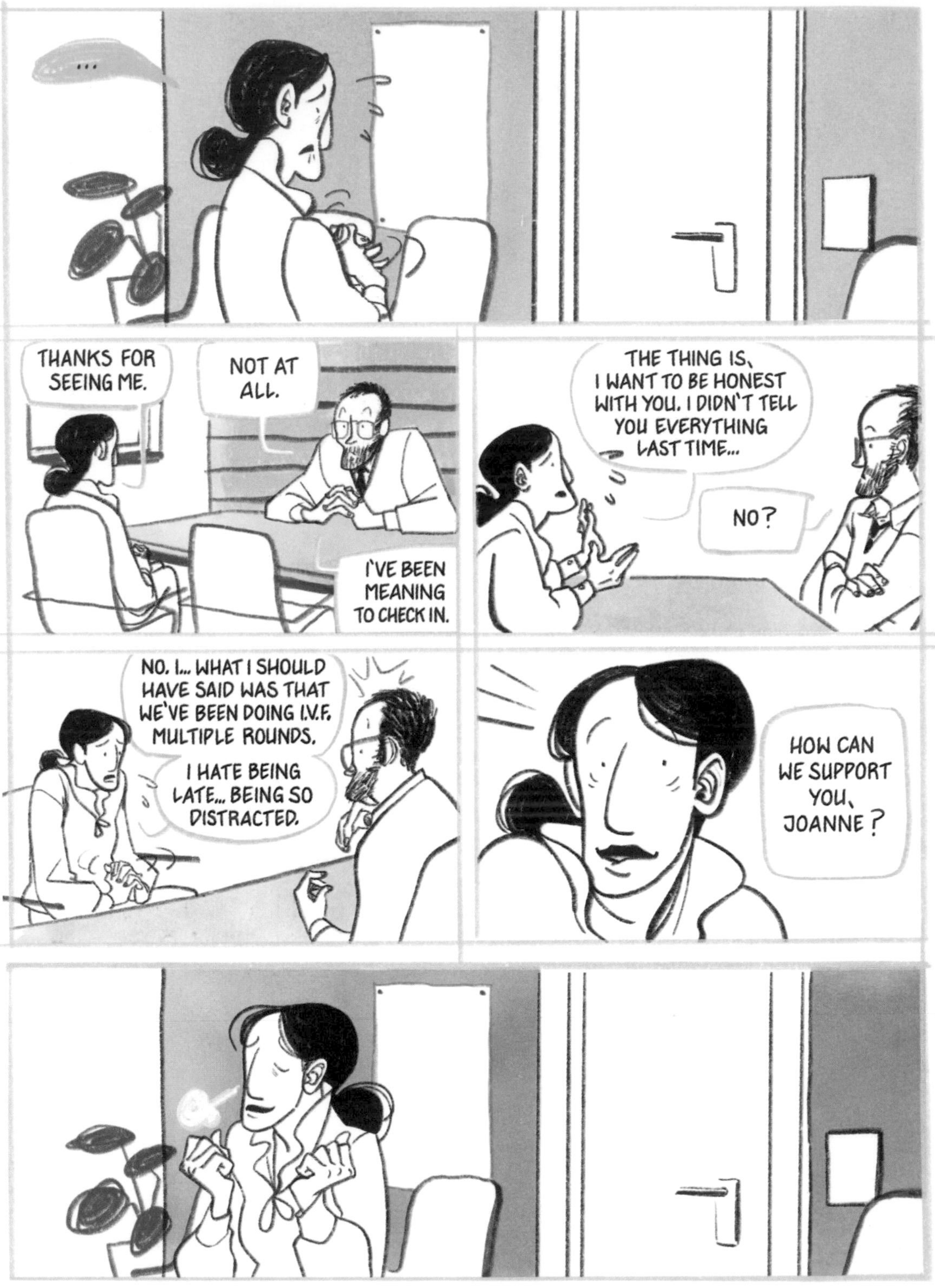
...
THANKS FOR SEEING ME.
NOT AT ALL.
I'VE BEEN MEANING TO CHECK IN.
THE THING IS, I WANT TO BE HONEST WITH YOU. I DIDN'T TELL YOU EVERYTHING LAST TIME...
NO?
NO. I... WHAT I SHOULD HAVE SAID WAS THAT WE'VE BEEN DOING I.V.F. MULTIPLE ROUNDS.
I HATE BEING LATE... BEING SO DISTRACTED.
HOW CAN WE SUPPORT YOU, JOANNE?

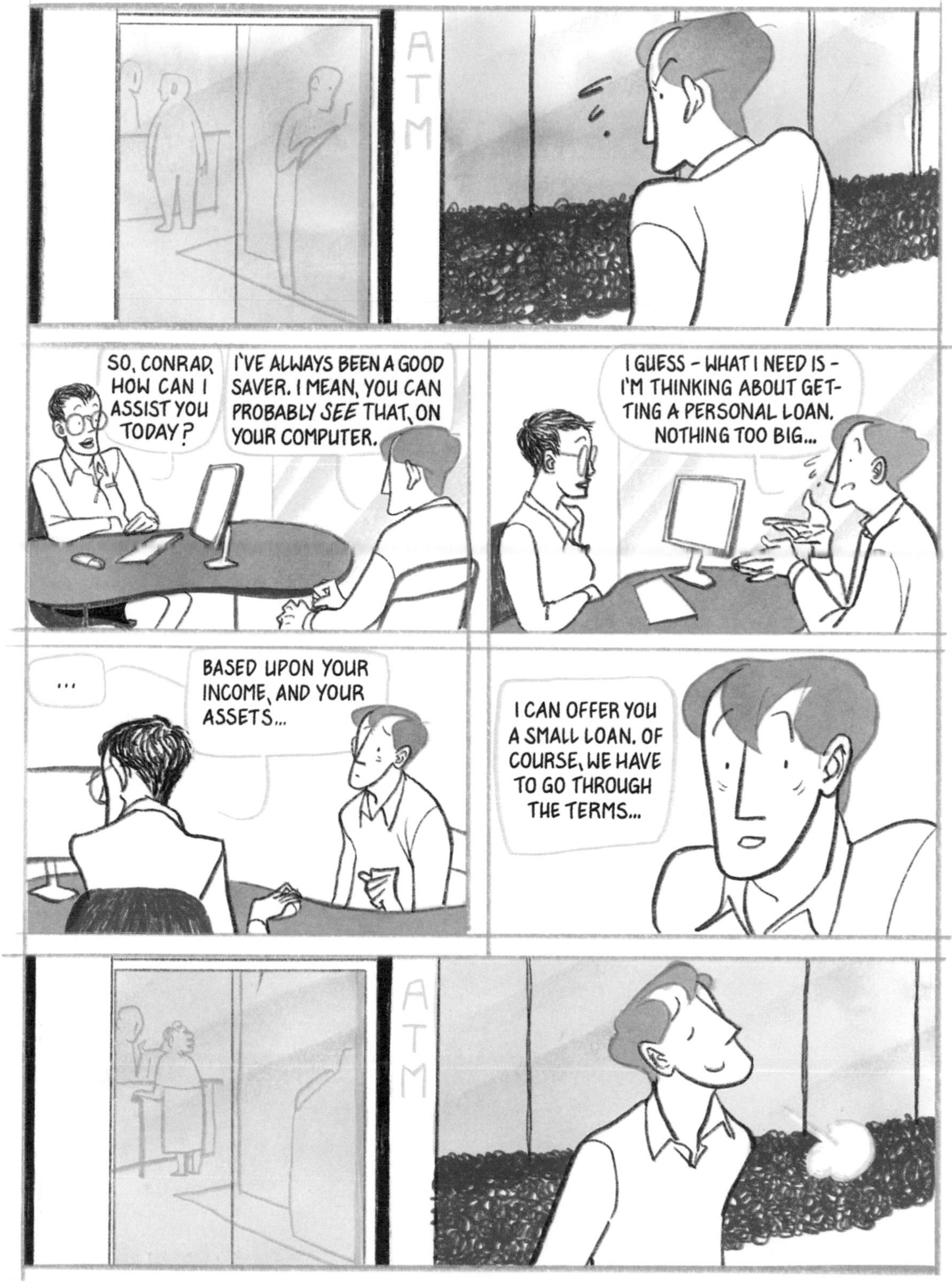

A T M
SO, CONRAD, HOW CAN I ASSIST YOU TODAY?
I'VE ALWAYS BEEN A GOOD SAVER. I MEAN, YOU CAN PROBABLY *SEE* THAT, ON YOUR COMPUTER.
I GUESS - WHAT I NEED IS - I'M THINKING ABOUT GETTING A PERSONAL LOAN. NOTHING TOO BIG...
...
BASED UPON YOUR INCOME, AND YOUR ASSETS...
I CAN OFFER YOU A SMALL LOAN. OF COURSE, WE HAVE TO GO THROUGH THE TERMS...
A T M

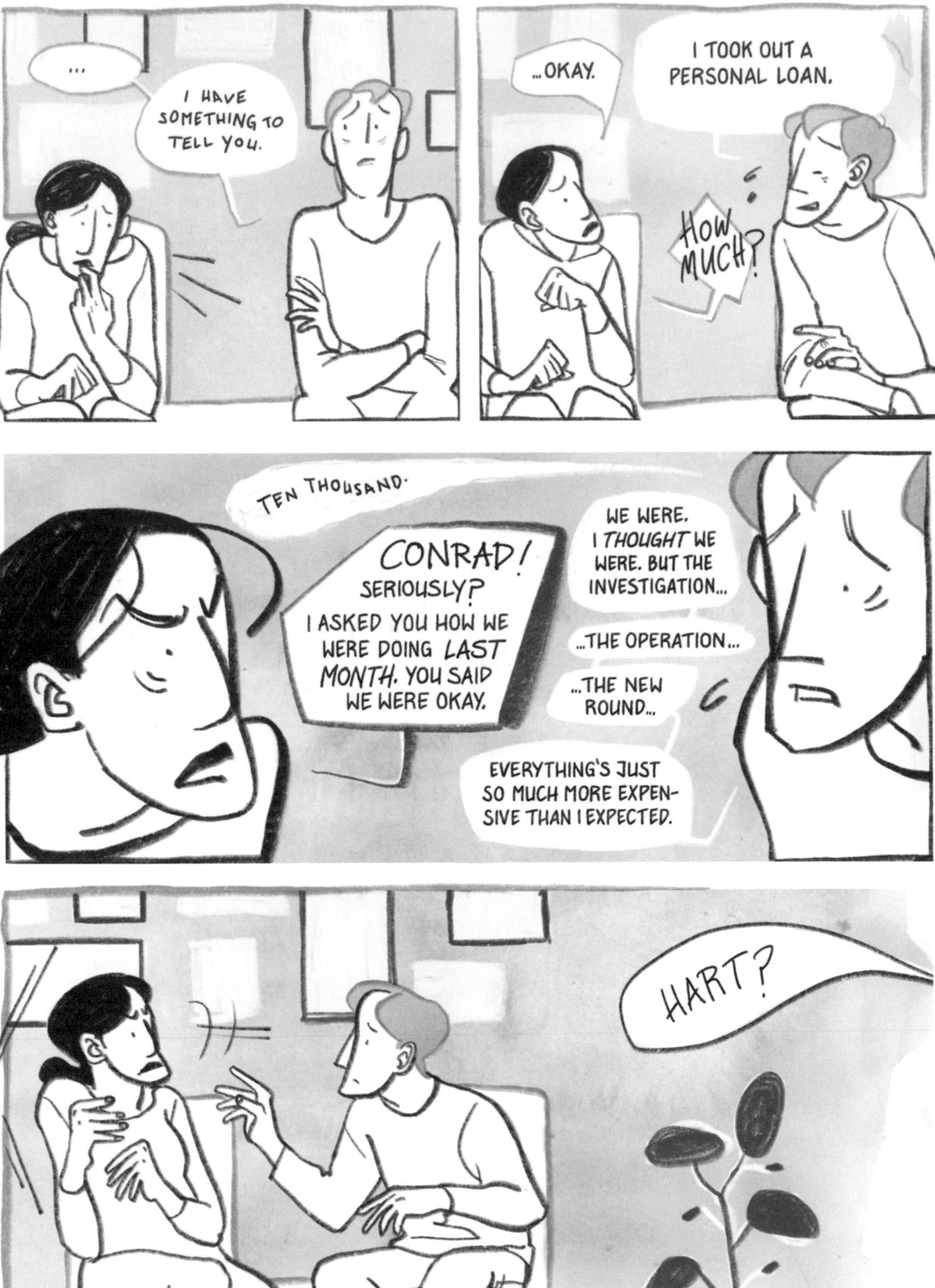
...
I HAVE SOMETHING TO TELL YOU.
...OKAY.
I TOOK OUT A PERSONAL LOAN.
HOW MUCH?
TEN THOUSAND.
CONRAD! SERIOUSLY? I ASKED YOU HOW WE WERE DOING LAST MONTH. YOU SAID WE WERE OKAY.
WE WERE. I THOUGHT WE WERE. BUT THE INVESTIGATION...
...THE OPERATION...
...THE NEW ROUND...
EVERYTHING'S JUST SO MUCH MORE EXPENSIVE THAN I EXPECTED.
HART?

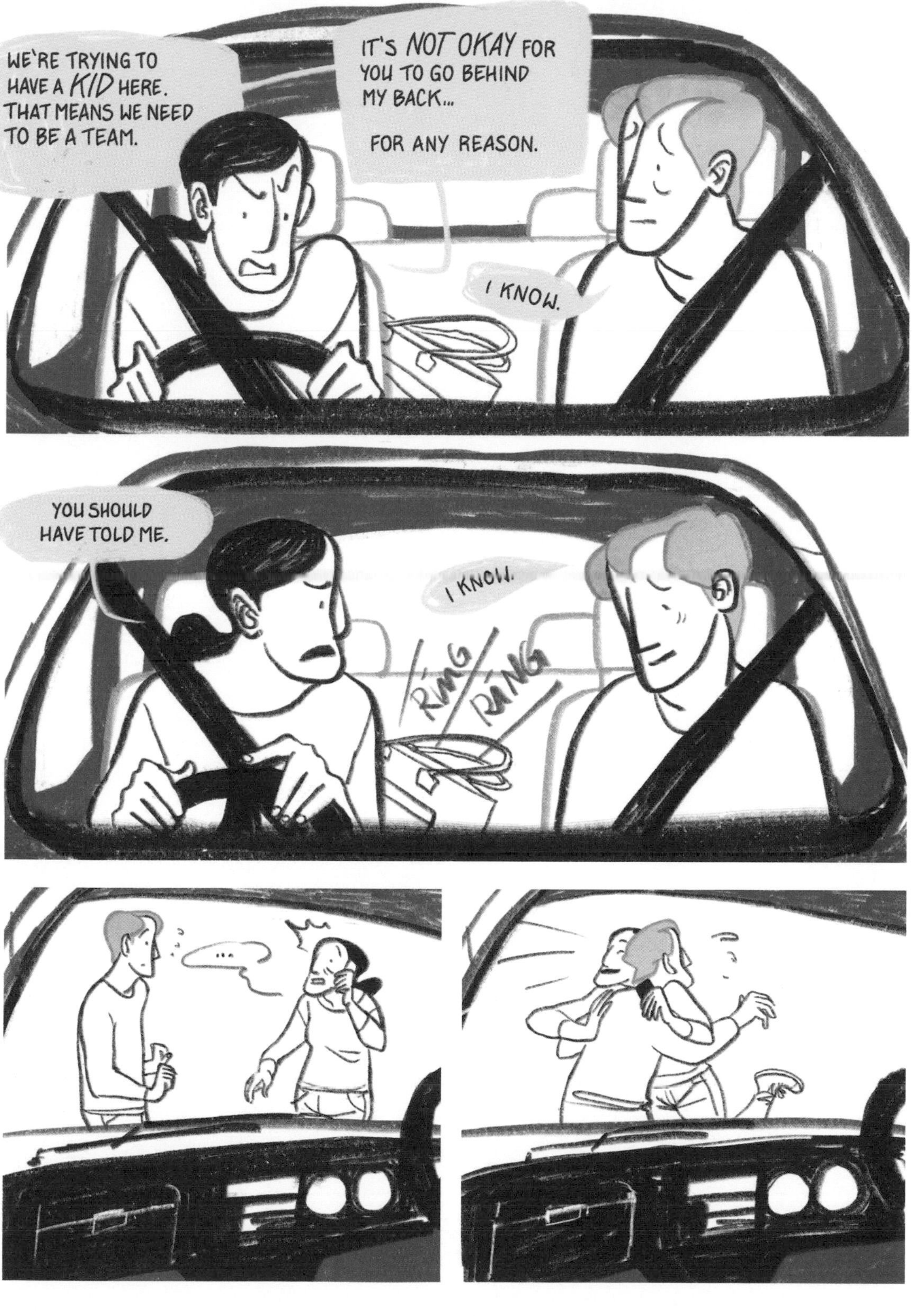
WE'RE TRYING TO HAVE A KID HERE. THAT MEANS WE NEED TO BE A TEAM.
IT'S NOT OKAY FOR YOU TO GO BEHIND MY BACK...
FOR ANY REASON.
I KNOW.
YOU SHOULD HAVE TOLD ME.
I KNOW.
RING RING
...

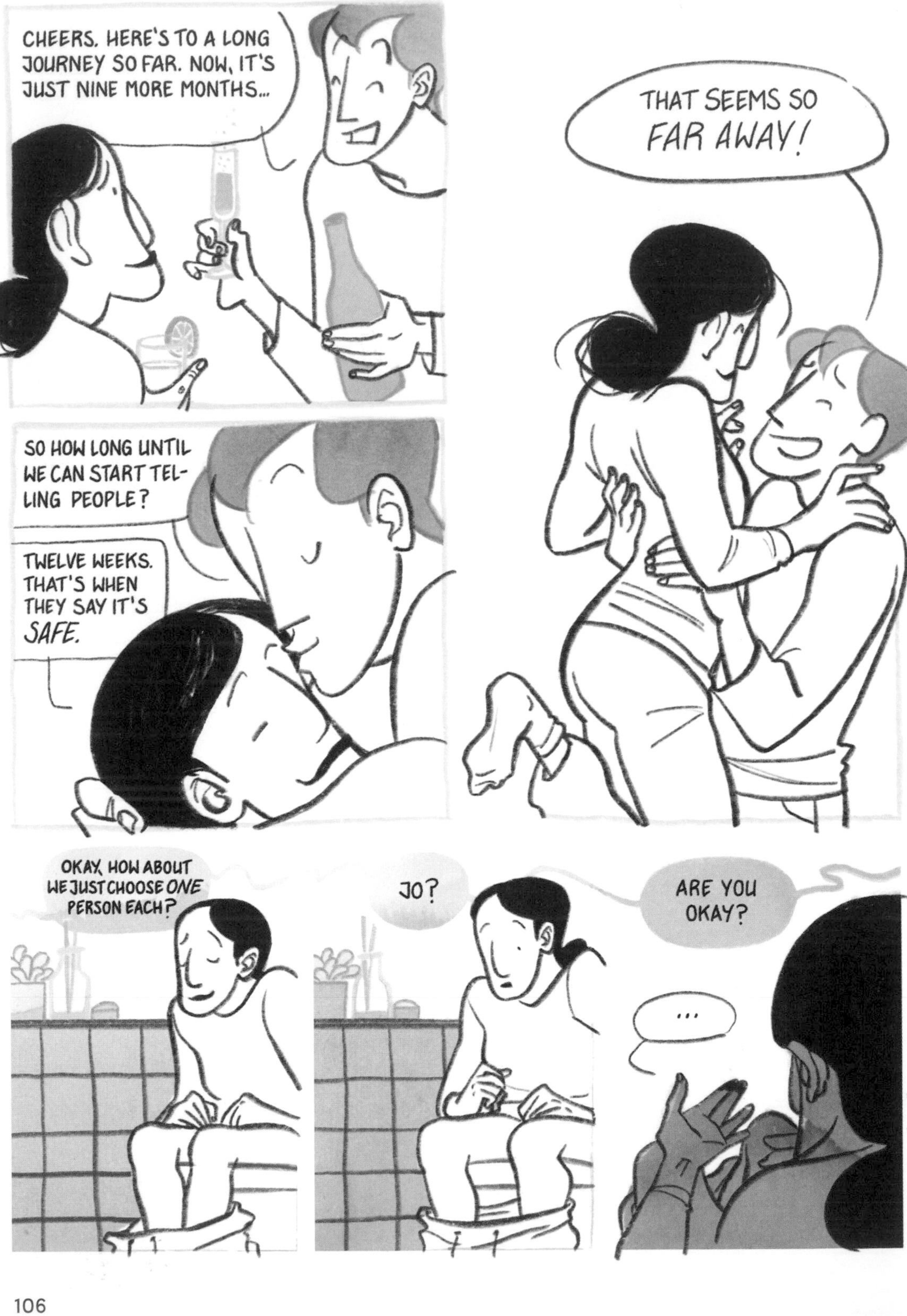
CHEERS. HERE'S TO A LONG JOURNEY SO FAR. NOW, IT'S JUST NINE MORE MONTHS...
THAT SEEMS SO FAR AWAY!
SO HOW LONG UNTIL WE CAN START TELLING PEOPLE?
TWELVE WEEKS. THAT'S WHEN THEY SAY IT'S SAFE.
OKAY, HOW ABOUT WE JUST CHOOSE ONE PERSON EACH?
JO?
ARE YOU OKAY?
...

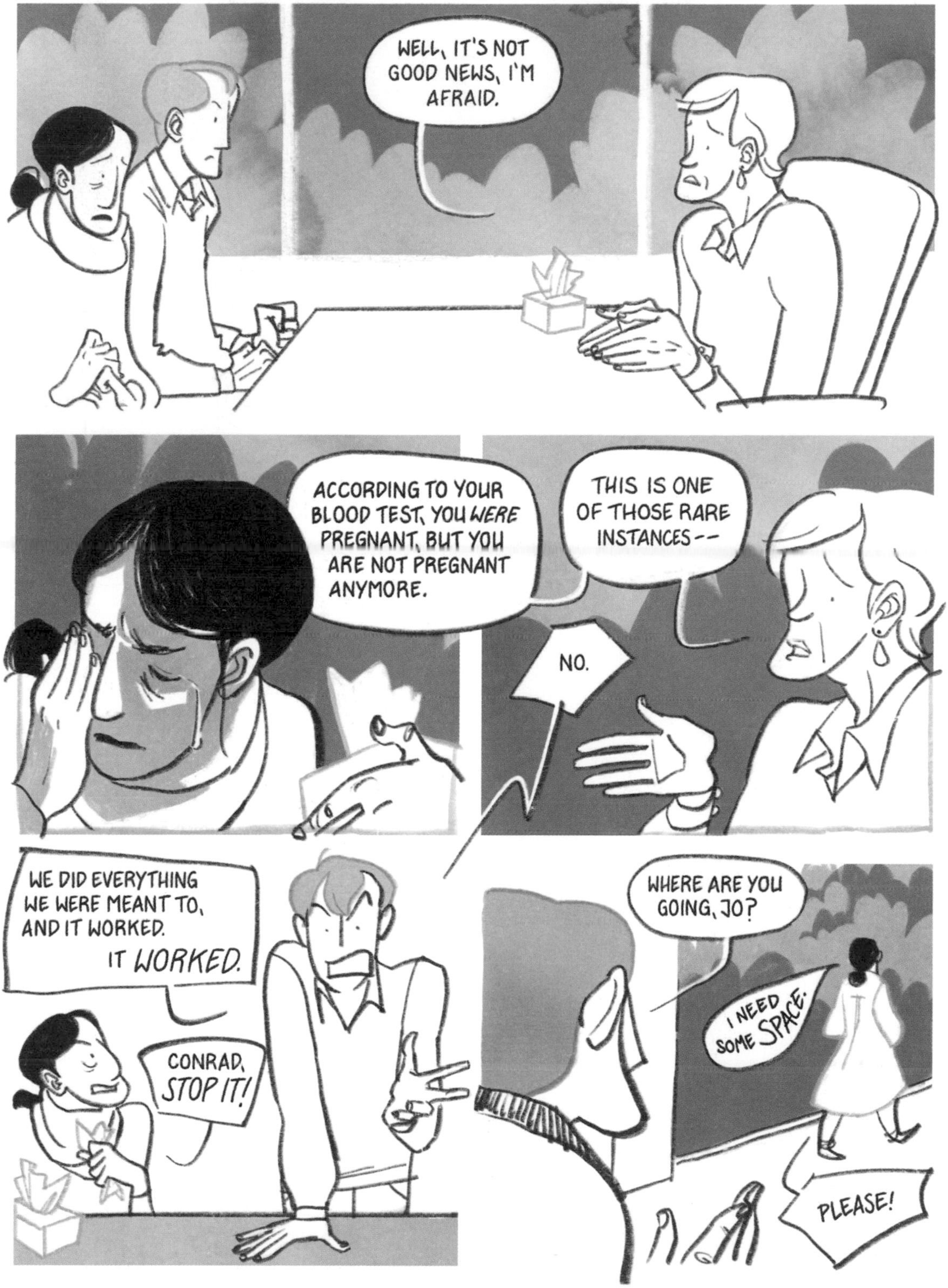
WELL, IT'S NOT GOOD NEWS, I'M AFRAID.
ACCORDING TO YOUR BLOOD TEST, YOU *WERE* PREGNANT, BUT YOU ARE NOT PREGNANT ANYMORE.
THIS IS ONE OF THOSE RARE INSTANCES--
NO.
WE DID EVERYTHING WE WERE MEANT TO, AND IT WORKED. IT *WORKED.*
CONRAD, *STOP IT!*
WHERE ARE YOU GOING, JO?
I NEED SOME *SPACE.*
PLEASE!

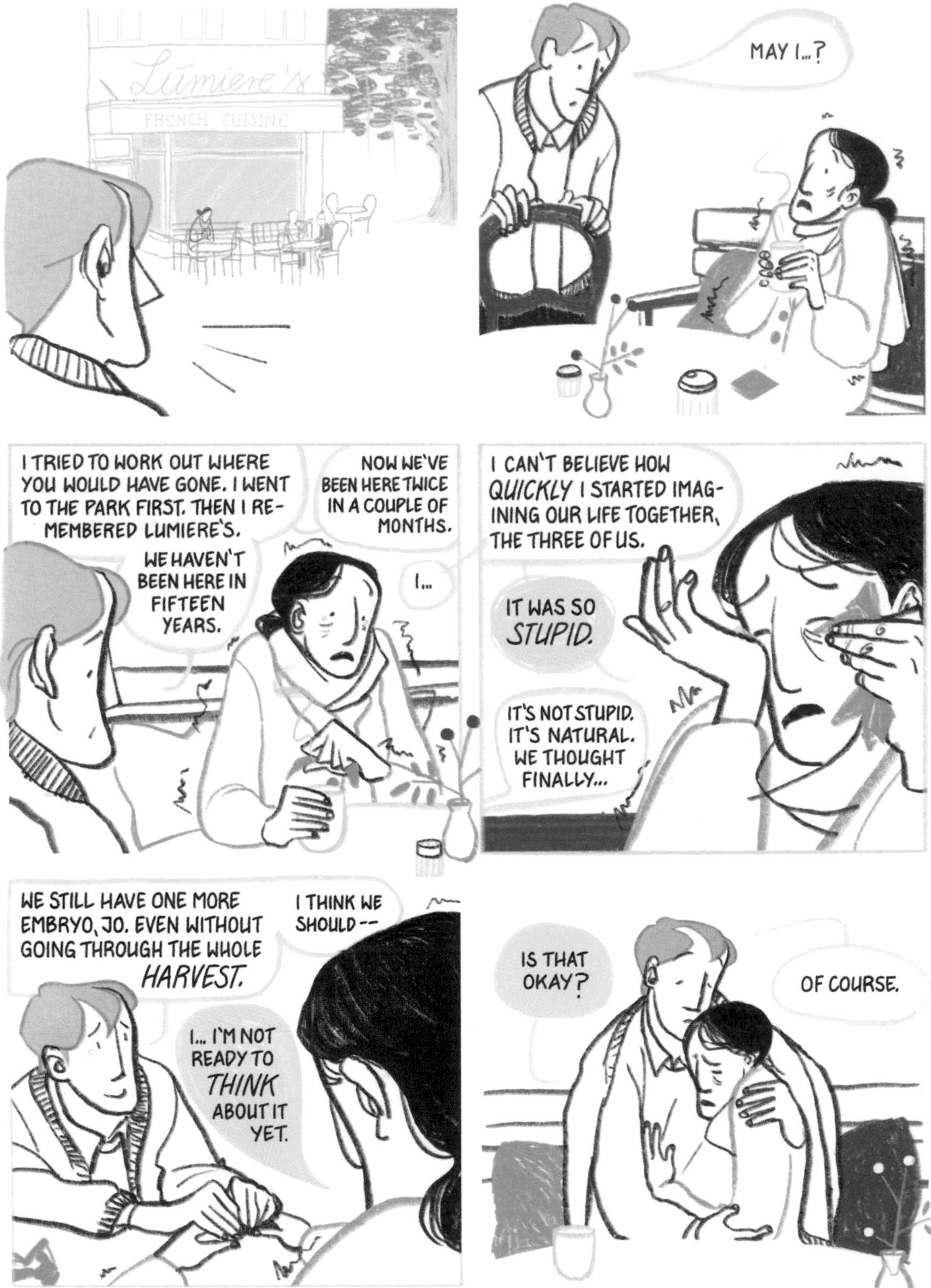
Lumiere's
FRENCH CUISINE
MAY I...?
I TRIED TO WORK OUT WHERE YOU WOULD HAVE GONE. I WENT TO THE PARK FIRST. THEN I REMEMBERED LUMIERE'S.
NOW WE'VE BEEN HERE TWICE IN A COUPLE OF MONTHS.
WE HAVEN'T BEEN HERE IN FIFTEEN YEARS.
I...
I CAN'T BELIEVE HOW QUICKLY I STARTED IMAGINING OUR LIFE TOGETHER, THE THREE OF US.
IT WAS SO STUPID.
IT'S NOT STUPID. IT'S NATURAL. WE THOUGHT FINALLY...
WE STILL HAVE ONE MORE EMBRYO, JO. EVEN WITHOUT GOING THROUGH THE WHOLE HARVEST.
I THINK WE SHOULD--
I... I'M NOT READY TO THINK ABOUT IT YET.
IS THAT OKAY?
OF COURSE.

CHAPTER 6:

THE GETAWAY

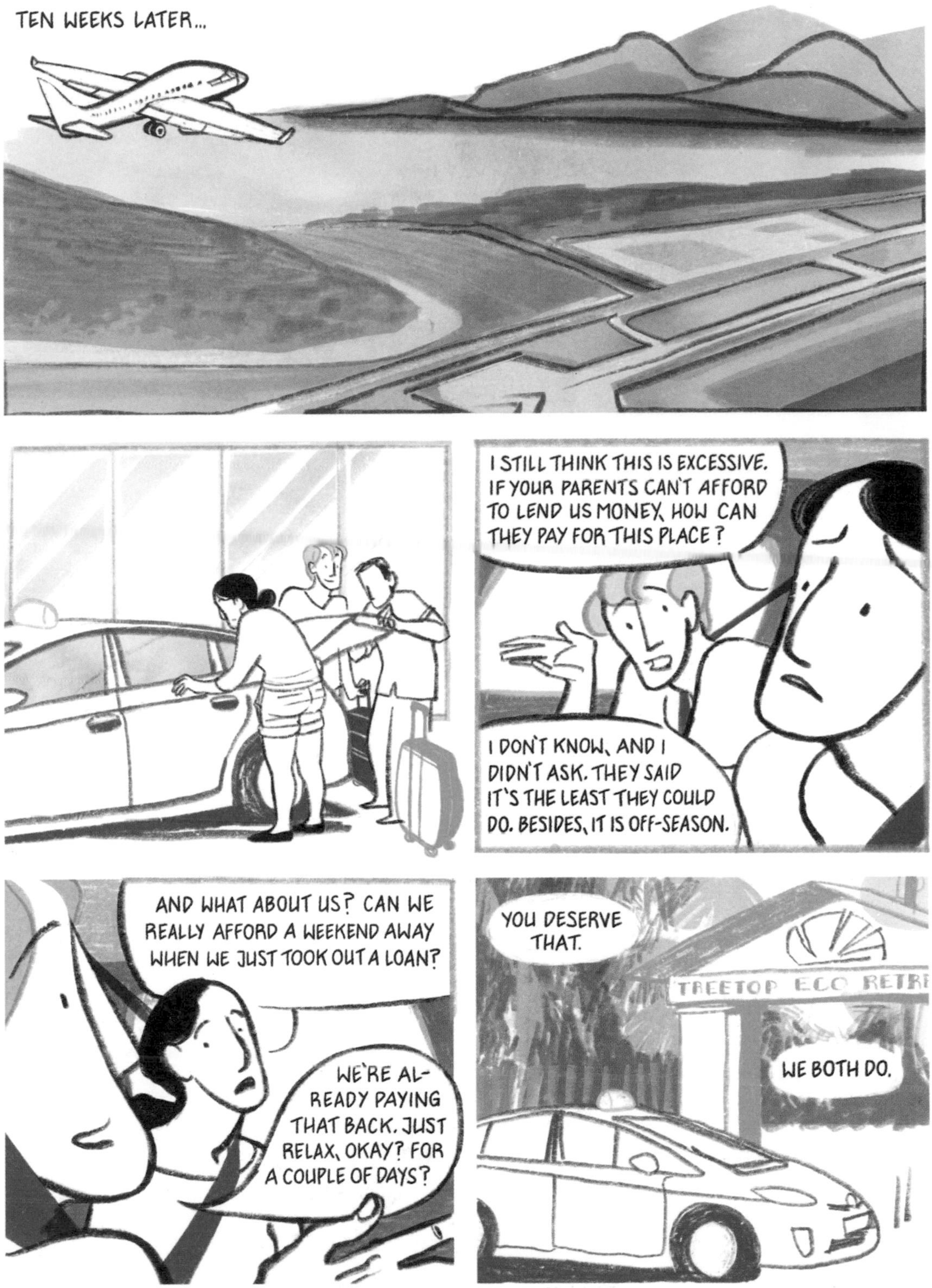
TEN WEEKS LATER...
I STILL THINK THIS IS EXCESSIVE. IF YOUR PARENTS CAN'T AFFORD TO LEND US MONEY, HOW CAN THEY PAY FOR THIS PLACE?
I DON'T KNOW, AND I DIDN'T ASK. THEY SAID IT'S THE LEAST THEY COULD DO. BESIDES, IT IS OFF-SEASON.
AND WHAT ABOUT US? CAN WE REALLY AFFORD A WEEKEND AWAY WHEN WE JUST TOOK OUT A LOAN?
WE'RE AL-READY PAYING THAT BACK. JUST RELAX, OKAY? FOR A COUPLE OF DAYS?
YOU DESERVE THAT.
TREETOP ECO RETR
WE BOTH DO.

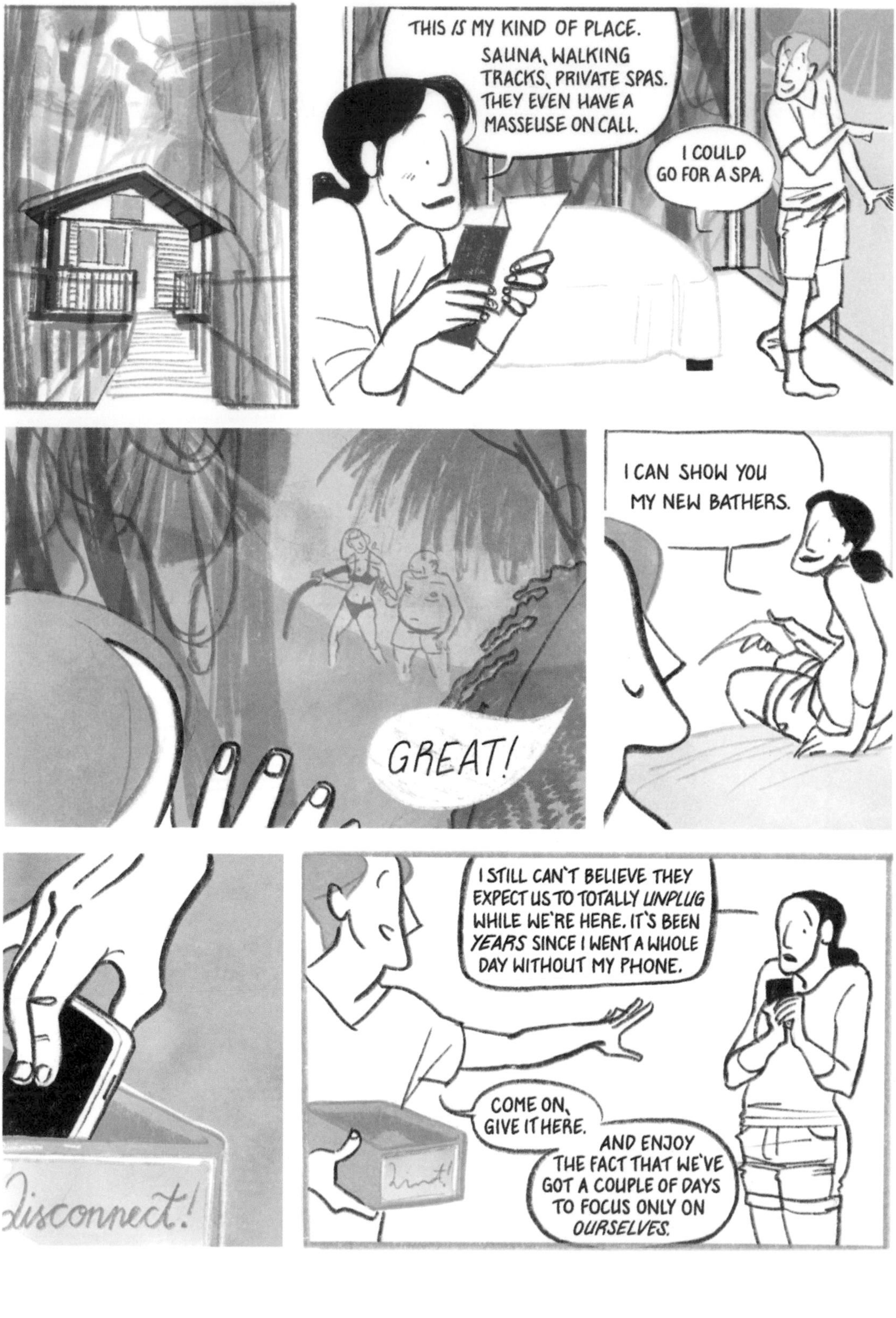

THIS *IS* MY KIND OF PLACE.
SAUNA, WALKING TRACKS, PRIVATE SPAS. THEY EVEN HAVE A MASSEUSE ON CALL.
I COULD GO FOR A SPA.
GREAT!
I CAN SHOW YOU MY NEW BATHERS.
Disconnect!
I STILL CAN'T BELIEVE THEY EXPECT US TO TOTALLY *UNPLUG* WHILE WE'RE HERE. IT'S BEEN *YEARS* SINCE I WENT A WHOLE DAY WITHOUT MY PHONE.
COME ON, GIVE IT HERE.
AND ENJOY THE FACT THAT WE'VE GOT A COUPLE OF DAYS TO FOCUS ONLY ON *OURSELVES*.

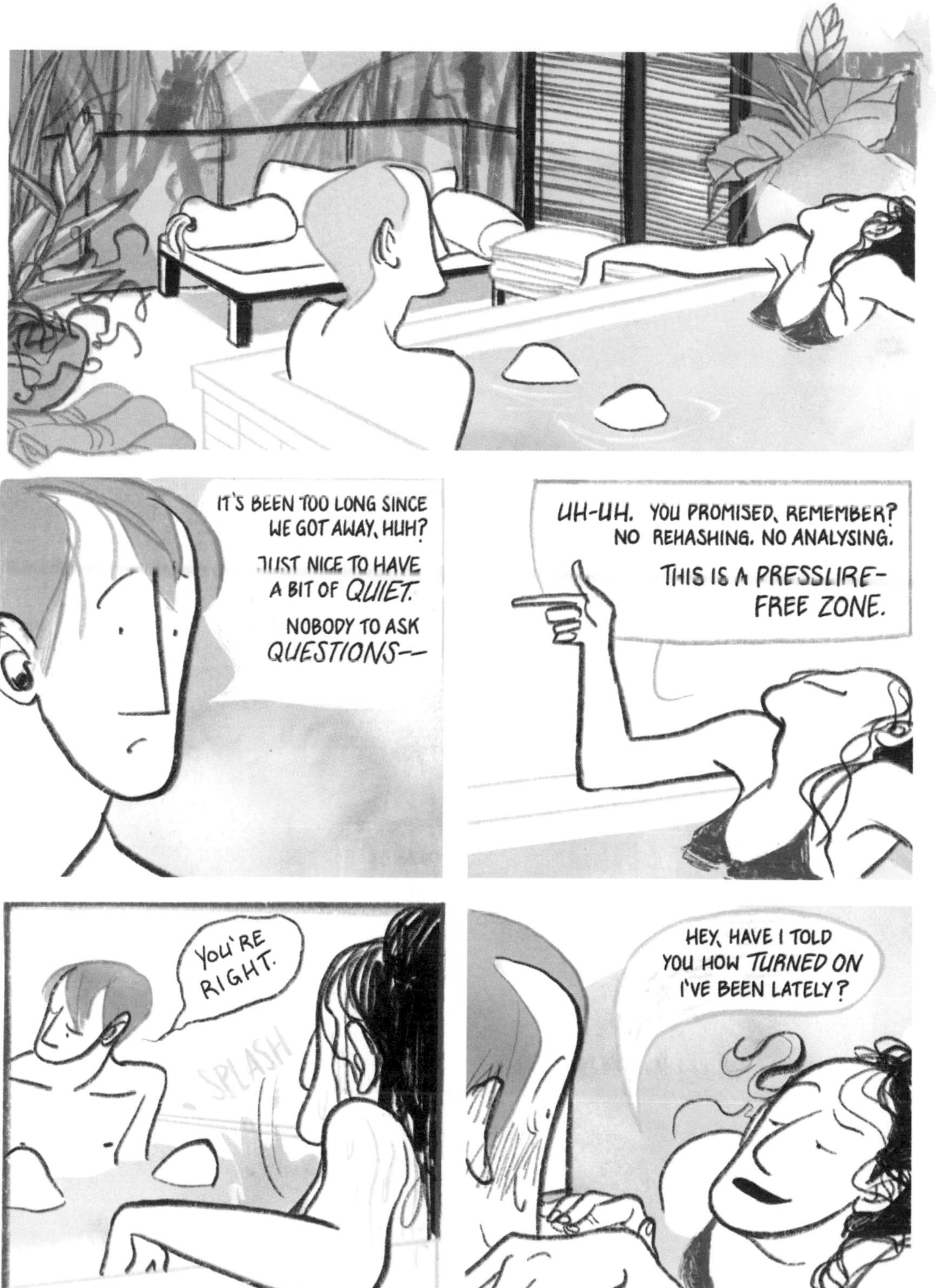
IT'S BEEN TOO LONG SINCE WE GOT AWAY, HUH?
JUST NICE TO HAVE A BIT OF QUIET.
NOBODY TO ASK QUESTIONS--
UH-UH. YOU PROMISED, REMEMBER? NO REHASHING. NO ANALYSING.
THIS IS A PRESSURE-FREE ZONE.
YOU'RE RIGHT.
SPLASH
HEY, HAVE I TOLD YOU HOW TURNED ON I'VE BEEN LATELY?

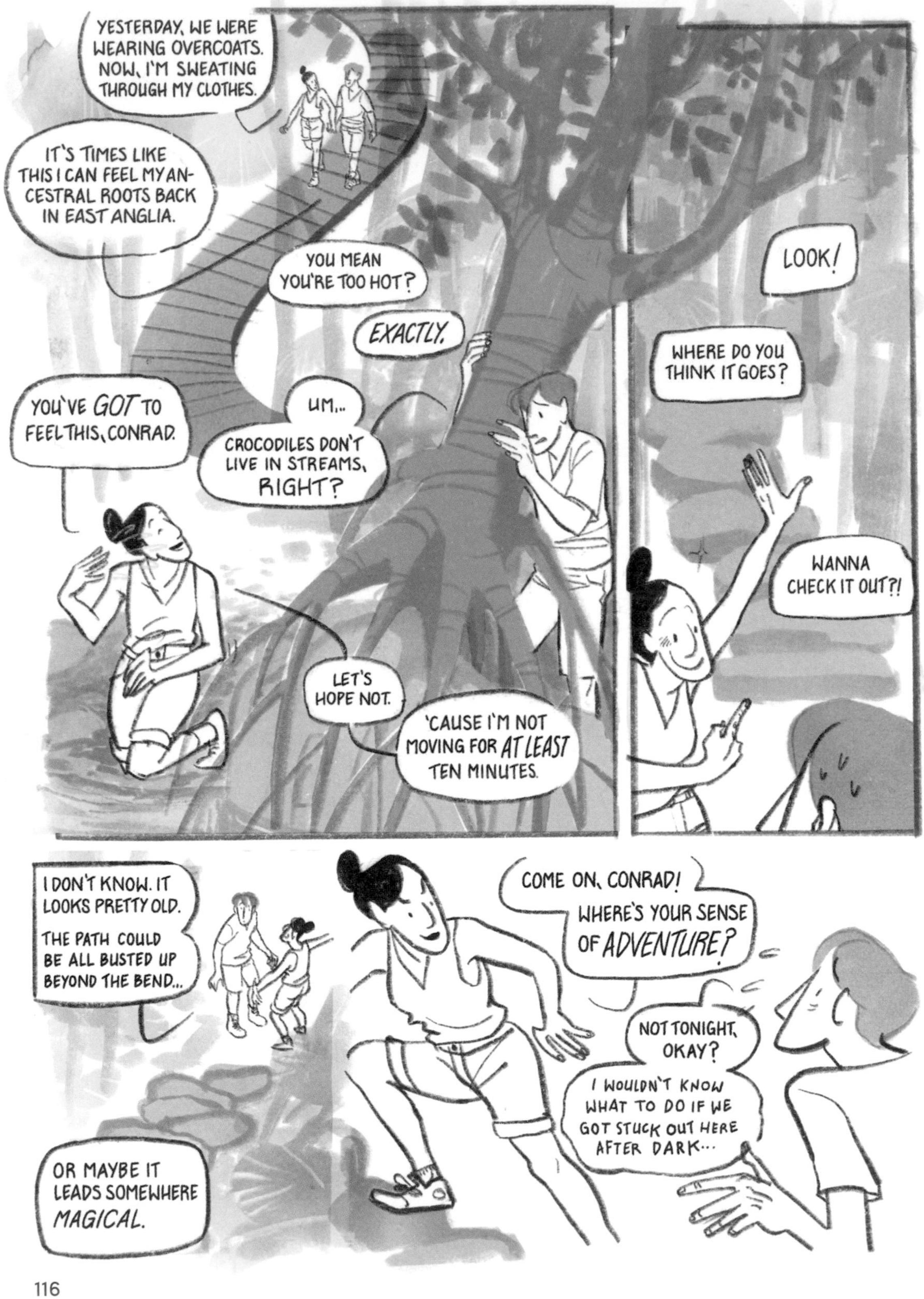
YESTERDAY, WE WERE WEARING OVERCOATS. NOW, I'M SWEATING THROUGH MY CLOTHES.
IT'S TIMES LIKE THIS I CAN FEEL MY ANCESTRAL ROOTS BACK IN EAST ANGLIA.
YOU MEAN YOU'RE TOO HOT?
EXACTLY.
YOU'VE GOT TO FEEL THIS, CONRAD.
UM...
CROCODILES DON'T LIVE IN STREAMS, RIGHT?
LET'S HOPE NOT.
'CAUSE I'M NOT MOVING FOR AT LEAST TEN MINUTES.
LOOK!
WHERE DO YOU THINK IT GOES?
WANNA CHECK IT OUT?!
I DON'T KNOW. IT LOOKS PRETTY OLD.
THE PATH COULD BE ALL BUSTED UP BEYOND THE BEND...
OR MAYBE IT LEADS SOMEWHERE MAGICAL.
COME ON, CONRAD!
WHERE'S YOUR SENSE OF ADVENTURE?
NOT TONIGHT, OKAY?
I WOULDN'T KNOW WHAT TO DO IF WE GOT STUCK OUT HERE AFTER DARK...

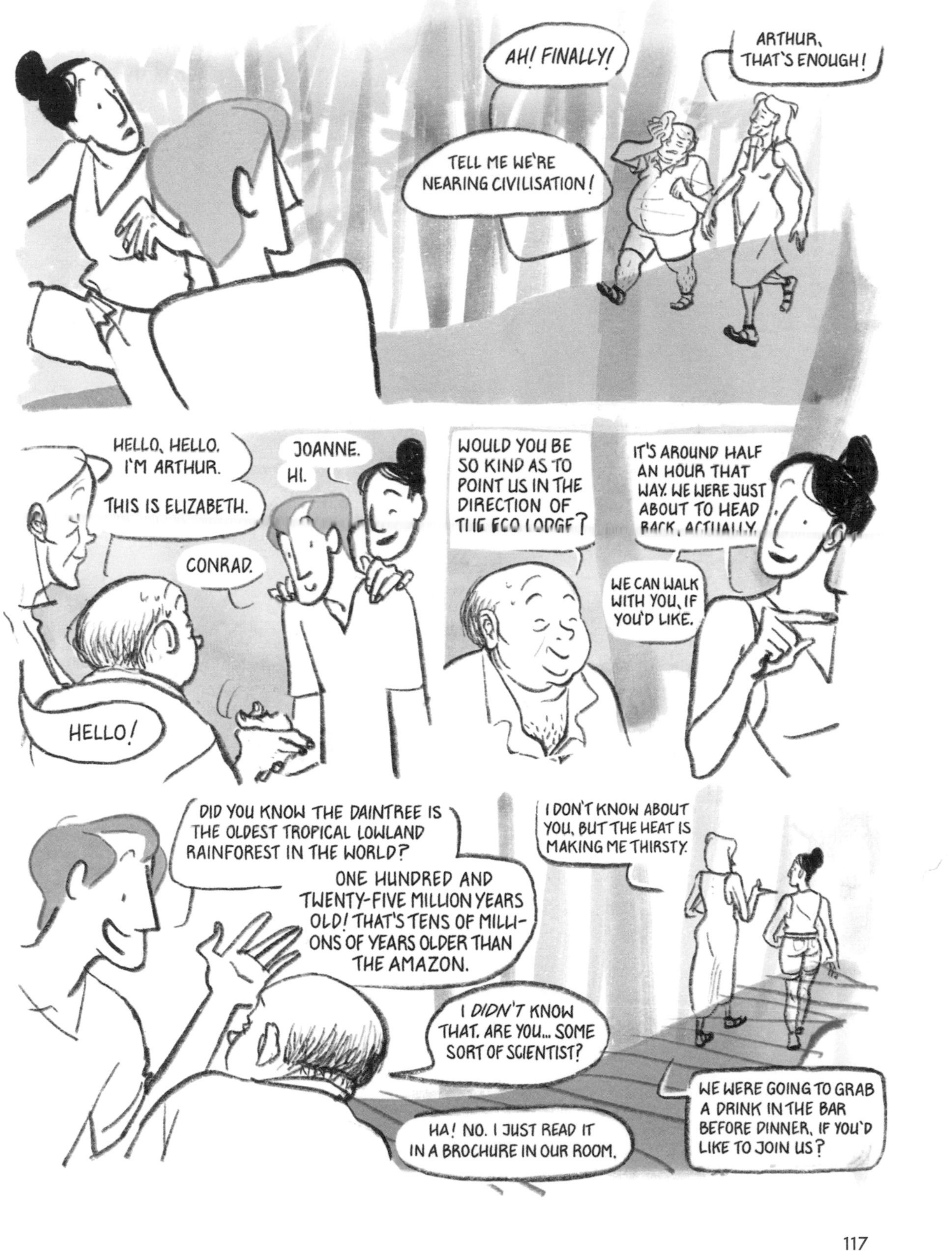
AH! FINALLY!
TELL ME WE'RE NEARING CIVILISATION!
ARTHUR, THAT'S ENOUGH!
HELLO, HELLO. I'M ARTHUR.
THIS IS ELIZABETH.
CONRAD.
HELLO!
JOANNE.
HI.
WOULD YOU BE SO KIND AS TO POINT US IN THE DIRECTION OF THE ECO LODGE?
IT'S AROUND HALF AN HOUR THAT WAY. WE WERE JUST ABOUT TO HEAD BACK, ACTUALLY.
WE CAN WALK WITH YOU, IF YOU'D LIKE.
DID YOU KNOW THE DAINTREE IS THE OLDEST TROPICAL LOWLAND RAINFOREST IN THE WORLD?
ONE HUNDRED AND TWENTY-FIVE MILLION YEARS OLD! THAT'S TENS OF MILLIONS OF YEARS OLDER THAN THE AMAZON.
I DIDN'T KNOW THAT. ARE YOU... SOME SORT OF SCIENTIST?
HA! NO. I JUST READ IT IN A BROCHURE IN OUR ROOM.
I DON'T KNOW ABOUT YOU, BUT THE HEAT IS MAKING ME THIRSTY.
WE WERE GOING TO GRAB A DRINK IN THE BAR BEFORE DINNER, IF YOU'D LIKE TO JOIN US?

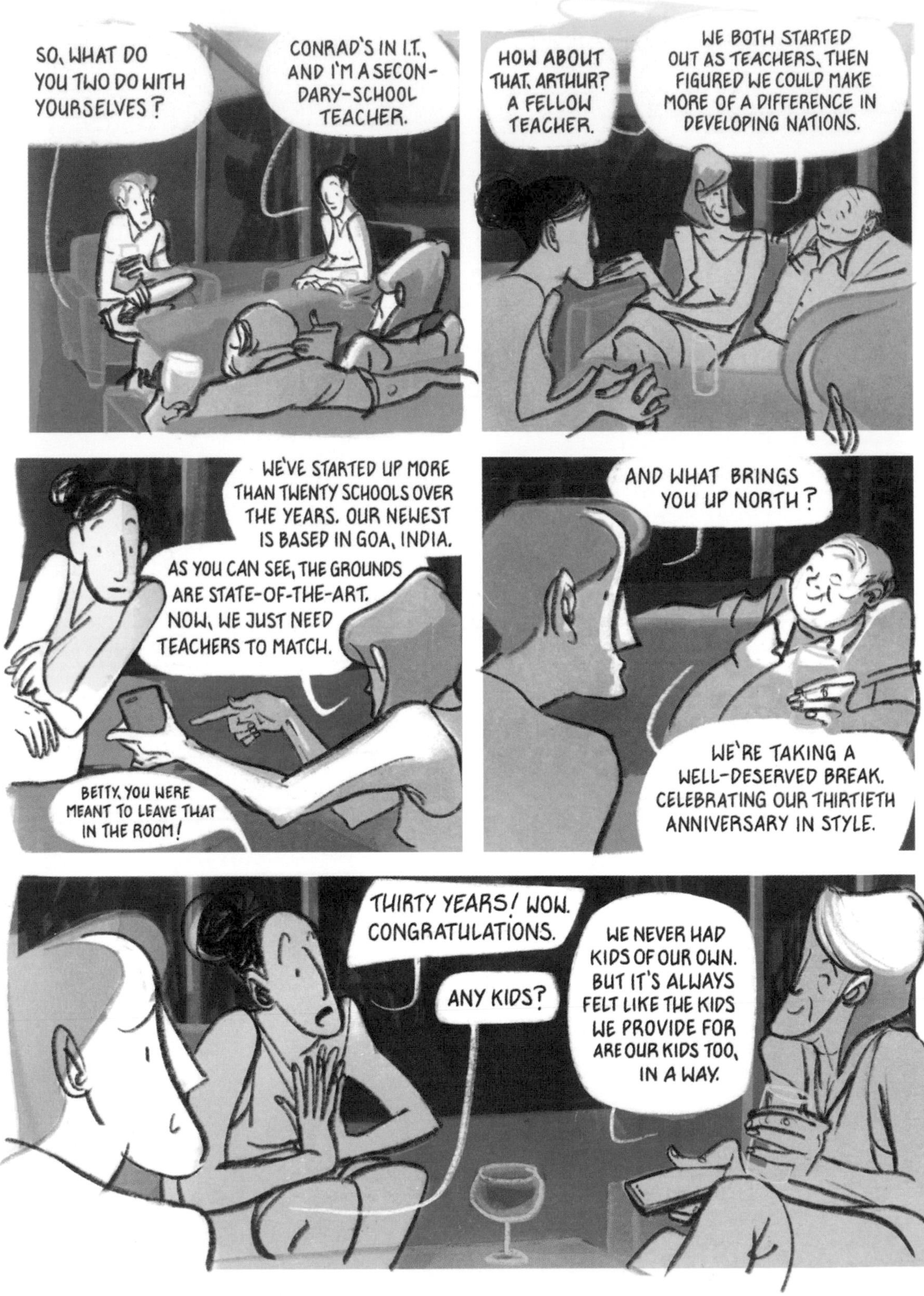
SO, WHAT DO YOU TWO DO WITH YOURSELVES?
CONRAD'S IN I.T., AND I'M A SECONDARY-SCHOOL TEACHER.
HOW ABOUT THAT, ARTHUR? A FELLOW TEACHER.
WE BOTH STARTED OUT AS TEACHERS, THEN FIGURED WE COULD MAKE MORE OF A DIFFERENCE IN DEVELOPING NATIONS.
WE'VE STARTED UP MORE THAN TWENTY SCHOOLS OVER THE YEARS. OUR NEWEST IS BASED IN GOA, INDIA.
AS YOU CAN SEE, THE GROUNDS ARE STATE-OF-THE-ART. NOW, WE JUST NEED TEACHERS TO MATCH.
BETTY, YOU WERE MEANT TO LEAVE THAT IN THE ROOM!
AND WHAT BRINGS YOU UP NORTH?
WE'RE TAKING A WELL-DESERVED BREAK. CELEBRATING OUR THIRTIETH ANNIVERSARY IN STYLE.
THIRTY YEARS! WOW. CONGRATULATIONS.
ANY KIDS?
WE NEVER HAD KIDS OF OUR OWN. BUT IT'S ALWAYS FELT LIKE THE KIDS WE PROVIDE FOR ARE OUR KIDS TOO, IN A WAY.

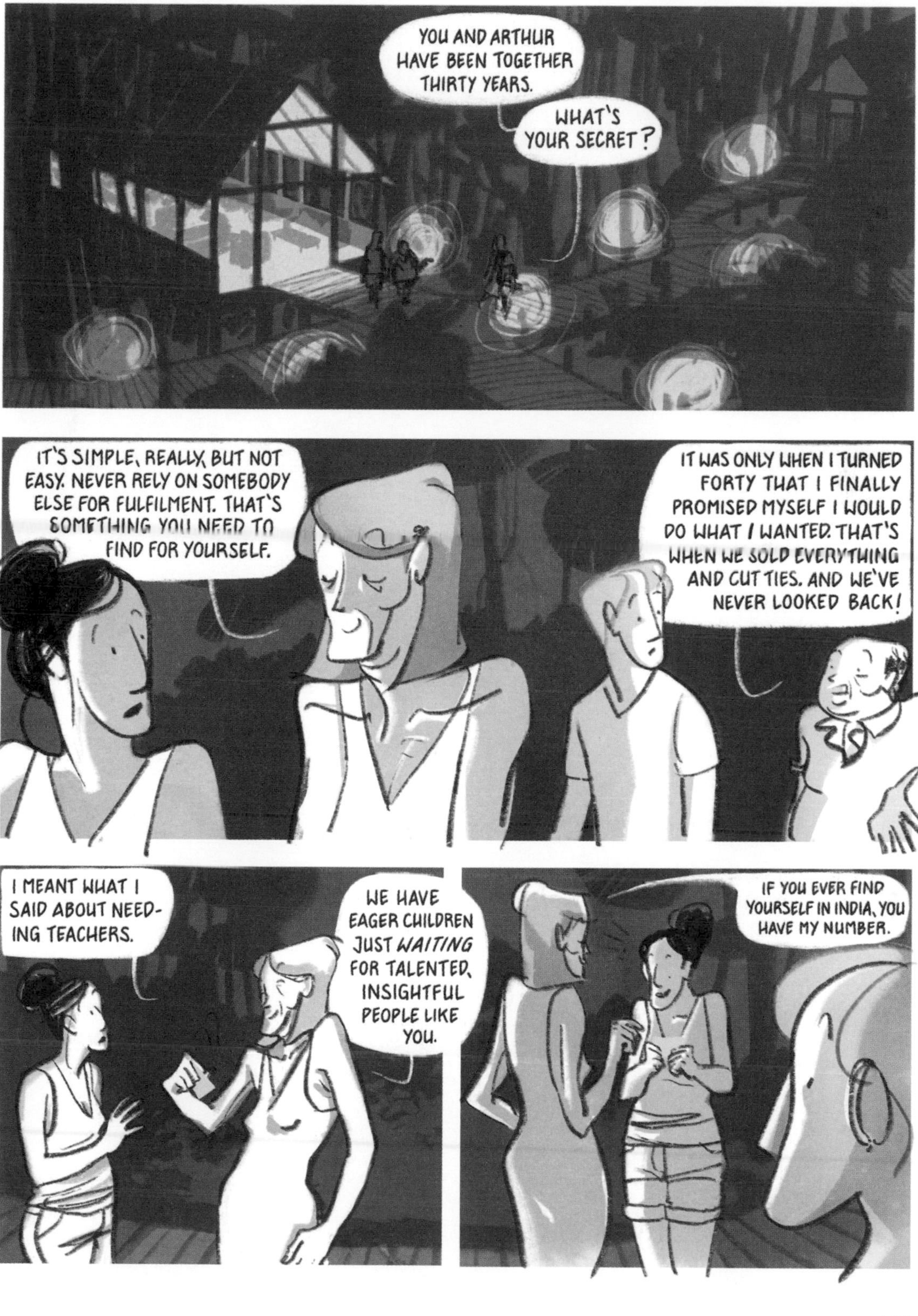
YOU AND ARTHUR HAVE BEEN TOGETHER THIRTY YEARS.
WHAT'S YOUR SECRET?
IT'S SIMPLE, REALLY, BUT NOT EASY. NEVER RELY ON SOMEBODY ELSE FOR FULFILMENT. THAT'S SOMETHING YOU NEED TO FIND FOR YOURSELF.
IT WAS ONLY WHEN I TURNED FORTY THAT I FINALLY PROMISED MYSELF I WOULD DO WHAT I WANTED. THAT'S WHEN WE SOLD EVERYTHING AND CUT TIES. AND WE'VE NEVER LOOKED BACK!
I MEANT WHAT I SAID ABOUT NEEDING TEACHERS.
WE HAVE EAGER CHILDREN JUST *WAITING* FOR TALENTED, INSIGHTFUL PEOPLE LIKE YOU.
IF YOU EVER FIND YOURSELF IN INDIA, YOU HAVE MY NUMBER.

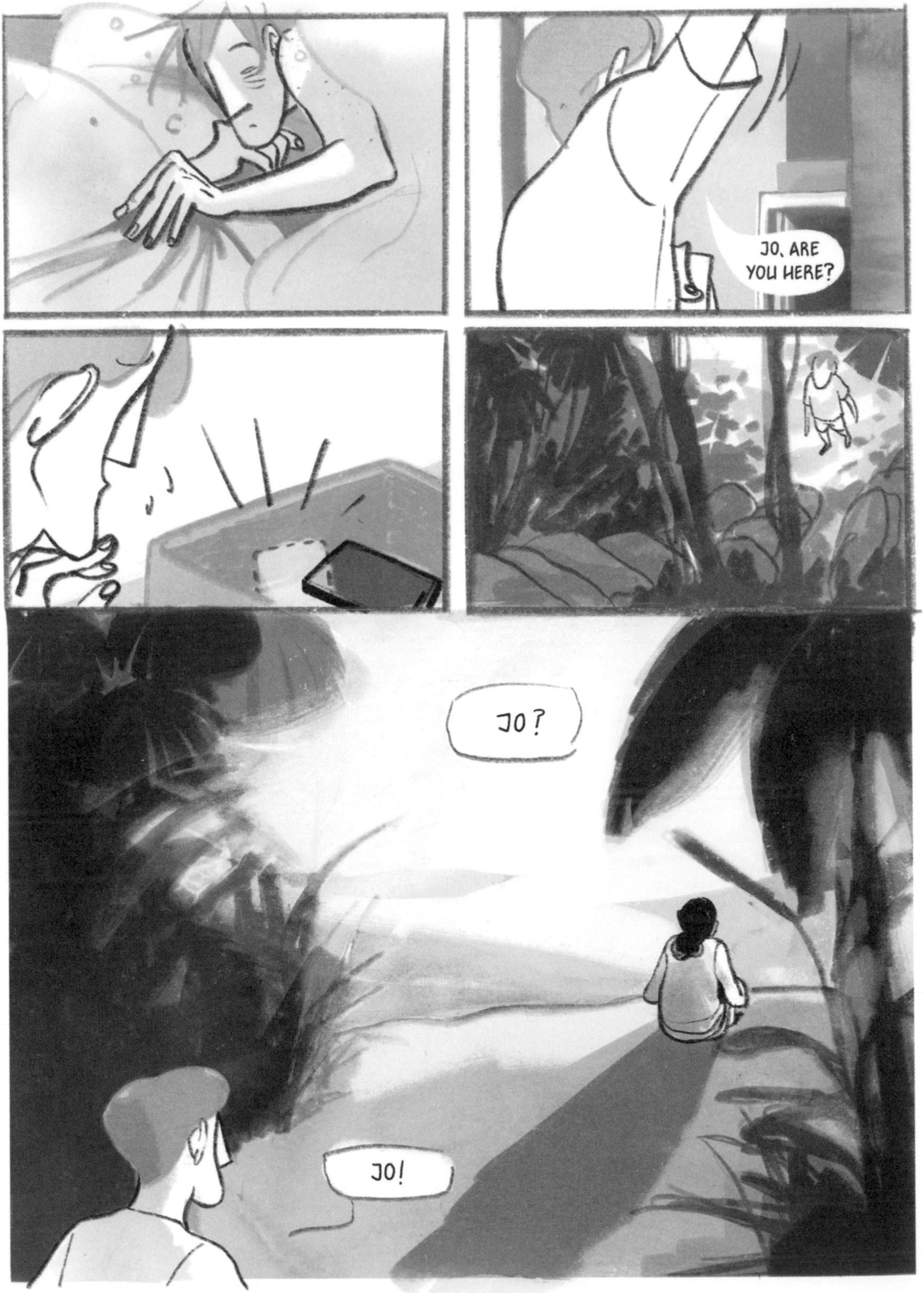
JO, ARE YOU HERE?
JO?
JO!

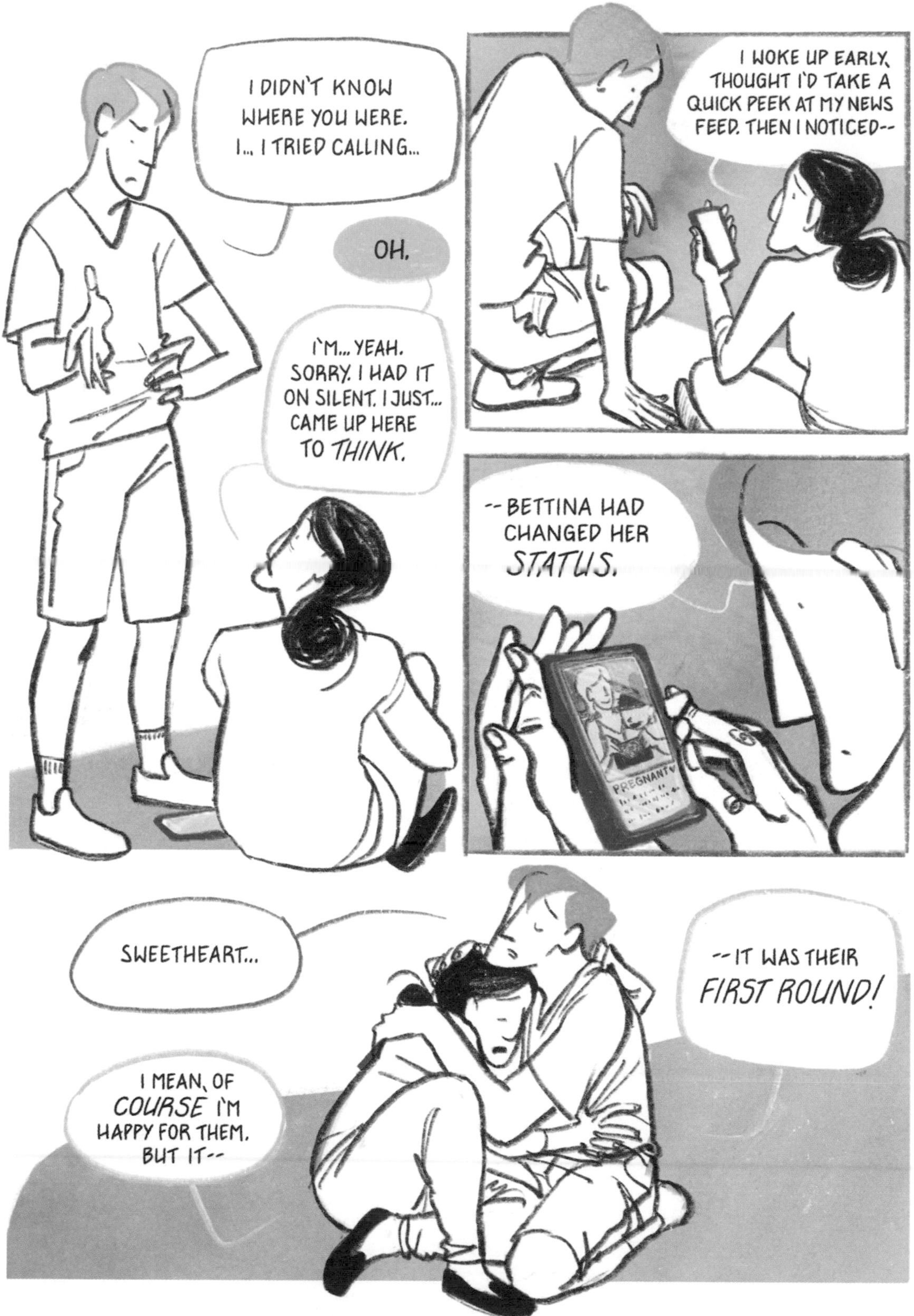
I DIDN'T KNOW WHERE YOU WERE. I... I TRIED CALLING...
OH.
I'M... YEAH. SORRY. I HAD IT ON SILENT. I JUST... CAME UP HERE TO THINK.
I WOKE UP EARLY, THOUGHT I'D TAKE A QUICK PEEK AT MY NEWS FEED. THEN I NOTICED--
--BETTINA HAD CHANGED HER STATUS.
PREGNANT
SWEETHEART...
--IT WAS THEIR FIRST ROUND!
I MEAN, OF COURSE I'M HAPPY FOR THEM. BUT IT--

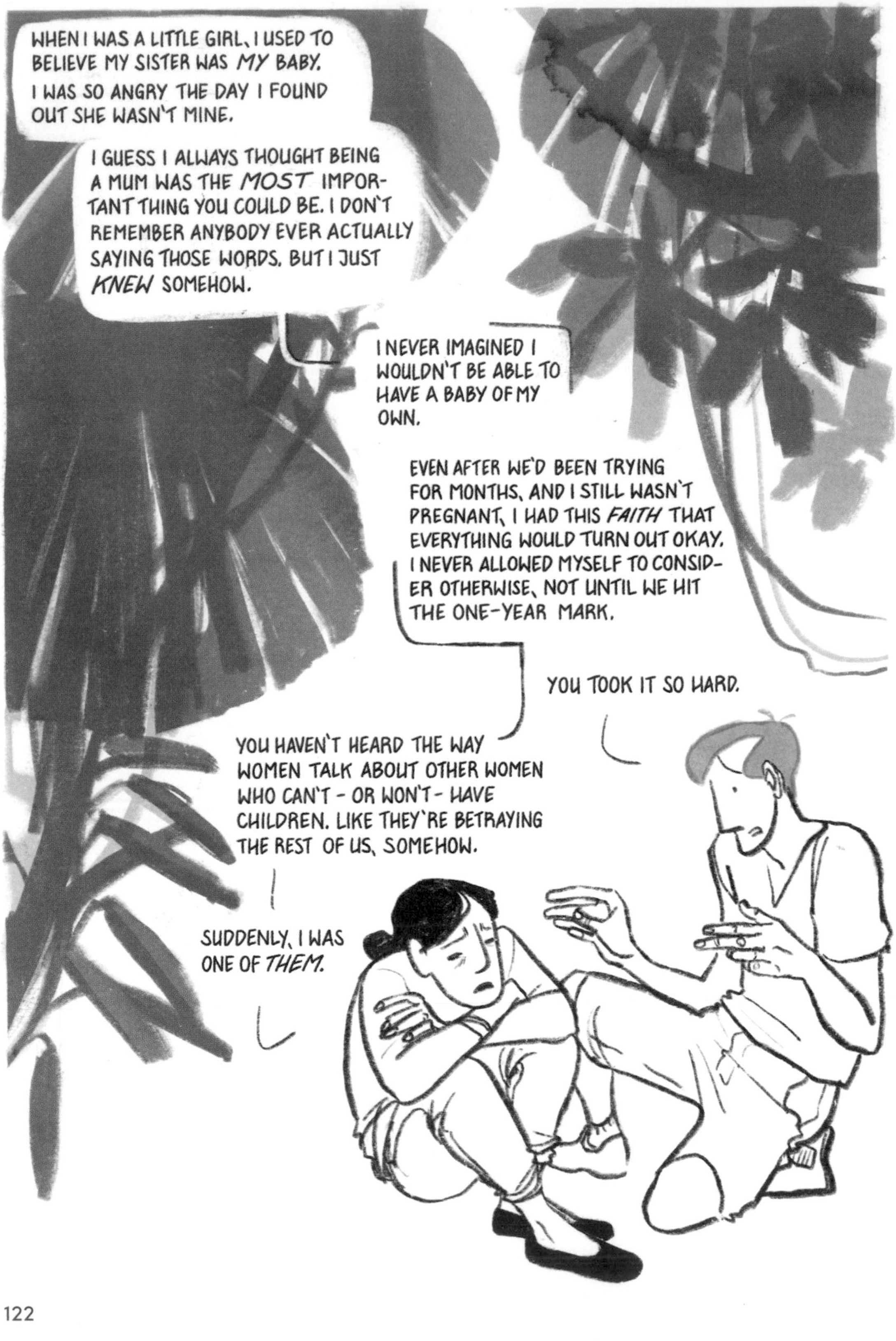
WHEN I WAS A LITTLE GIRL, I USED TO BELIEVE MY SISTER WAS MY BABY. I WAS SO ANGRY THE DAY I FOUND OUT SHE WASN'T MINE.
I GUESS I ALWAYS THOUGHT BEING A MUM WAS THE MOST IMPORTANT THING YOU COULD BE. I DON'T REMEMBER ANYBODY EVER ACTUALLY SAYING THOSE WORDS, BUT I JUST KNEW SOMEHOW.
I NEVER IMAGINED I WOULDN'T BE ABLE TO HAVE A BABY OF MY OWN.
EVEN AFTER WE'D BEEN TRYING FOR MONTHS, AND I STILL WASN'T PREGNANT, I HAD THIS FAITH THAT EVERYTHING WOULD TURN OUT OKAY. I NEVER ALLOWED MYSELF TO CONSIDER OTHERWISE, NOT UNTIL WE HIT THE ONE-YEAR MARK.
YOU TOOK IT SO HARD.
YOU HAVEN'T HEARD THE WAY WOMEN TALK ABOUT OTHER WOMEN WHO CAN'T - OR WON'T - HAVE CHILDREN. LIKE THEY'RE BETRAYING THE REST OF US, SOMEHOW.
SUDDENLY, I WAS ONE OF THEM.

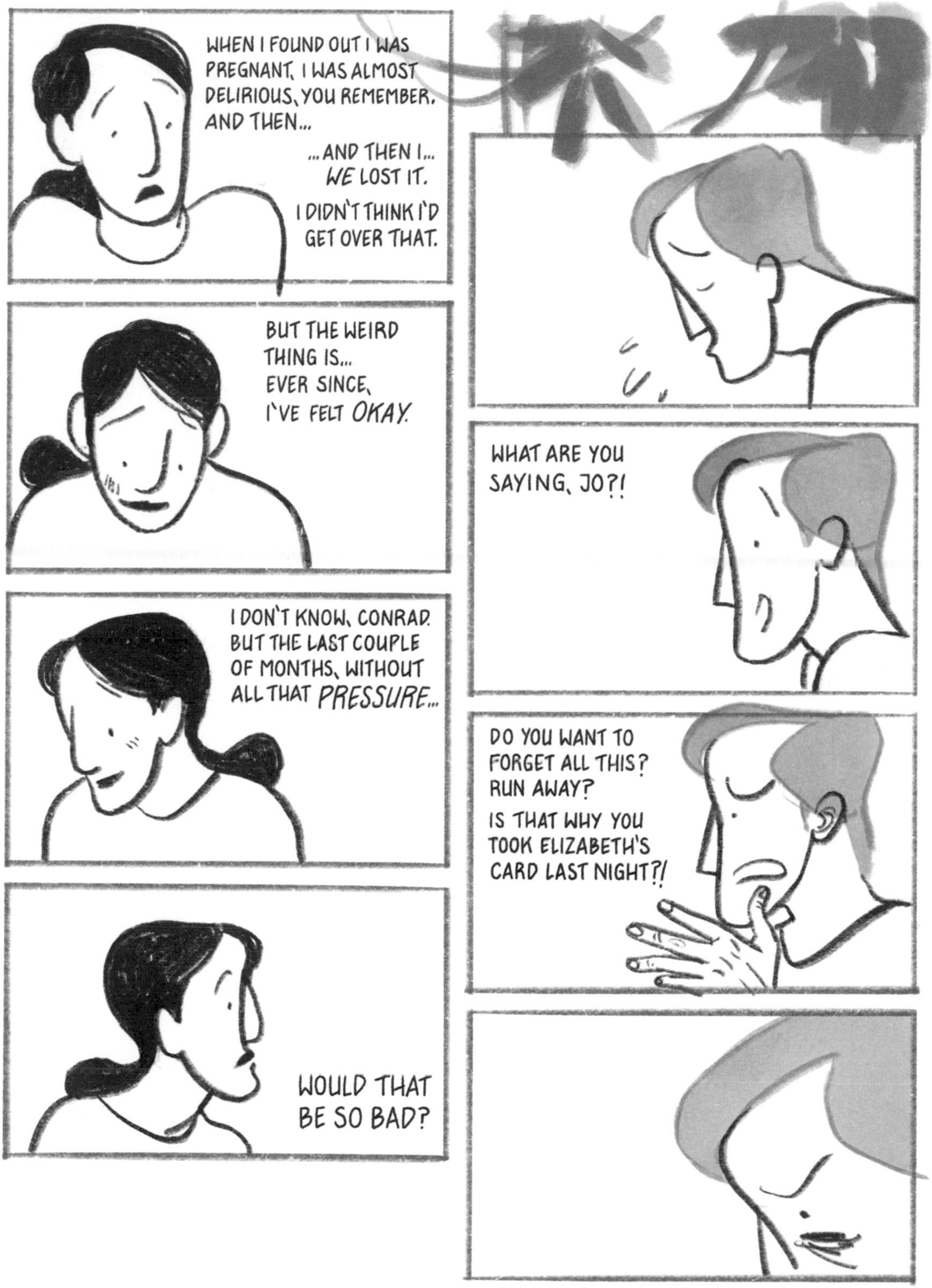
WHEN I FOUND OUT I WAS PREGNANT, I WAS ALMOST DELIRIOUS, YOU REMEMBER. AND THEN...
...AND THEN I... WE LOST IT.
I DIDN'T THINK I'D GET OVER THAT.
BUT THE WEIRD THING IS... EVER SINCE, I'VE FELT OKAY.
I DON'T KNOW, CONRAD. BUT THE LAST COUPLE OF MONTHS, WITHOUT ALL THAT PRESSURE...
WOULD THAT BE SO BAD?
WHAT ARE YOU SAYING, JO?!
DO YOU WANT TO FORGET ALL THIS? RUN AWAY?
IS THAT WHY YOU TOOK ELIZABETH'S CARD LAST NIGHT?!

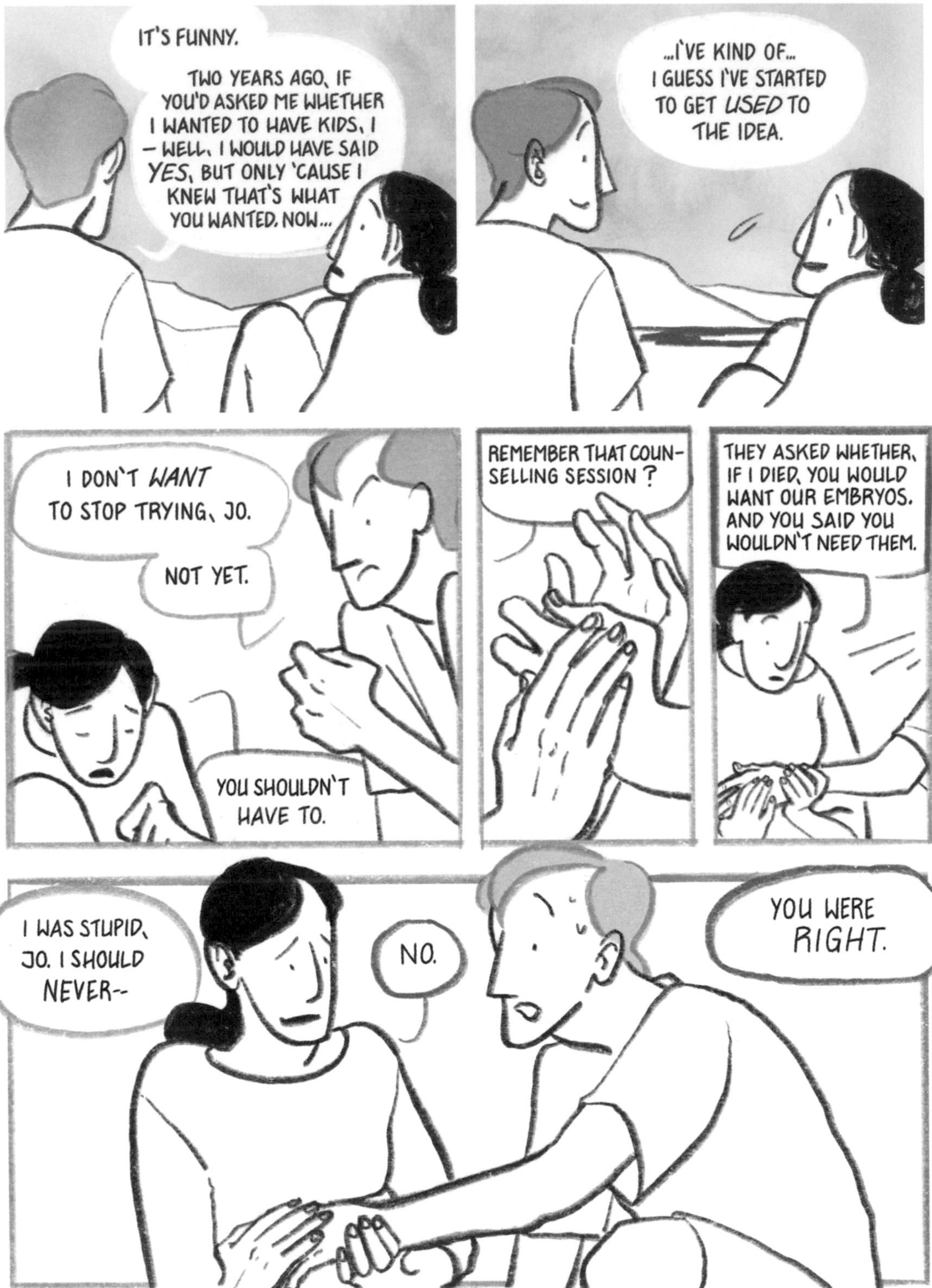
IT'S FUNNY.
TWO YEARS AGO, IF YOU'D ASKED ME WHETHER I WANTED TO HAVE KIDS, I – WELL, I WOULD HAVE SAID YES, BUT ONLY 'CAUSE I KNEW THAT'S WHAT YOU WANTED. NOW...
...I'VE KIND OF... I GUESS I'VE STARTED TO GET USED TO THE IDEA.
I DON'T WANT TO STOP TRYING, JO.
NOT YET.
YOU SHOULDN'T HAVE TO.
REMEMBER THAT COUN-SELLING SESSION?
THEY ASKED WHETHER, IF I DIED, YOU WOULD WANT OUR EMBRYOS. AND YOU SAID YOU WOULDN'T NEED THEM.
I WAS STUPID, JO. I SHOULD NEVER--
NO.
YOU WERE RIGHT.

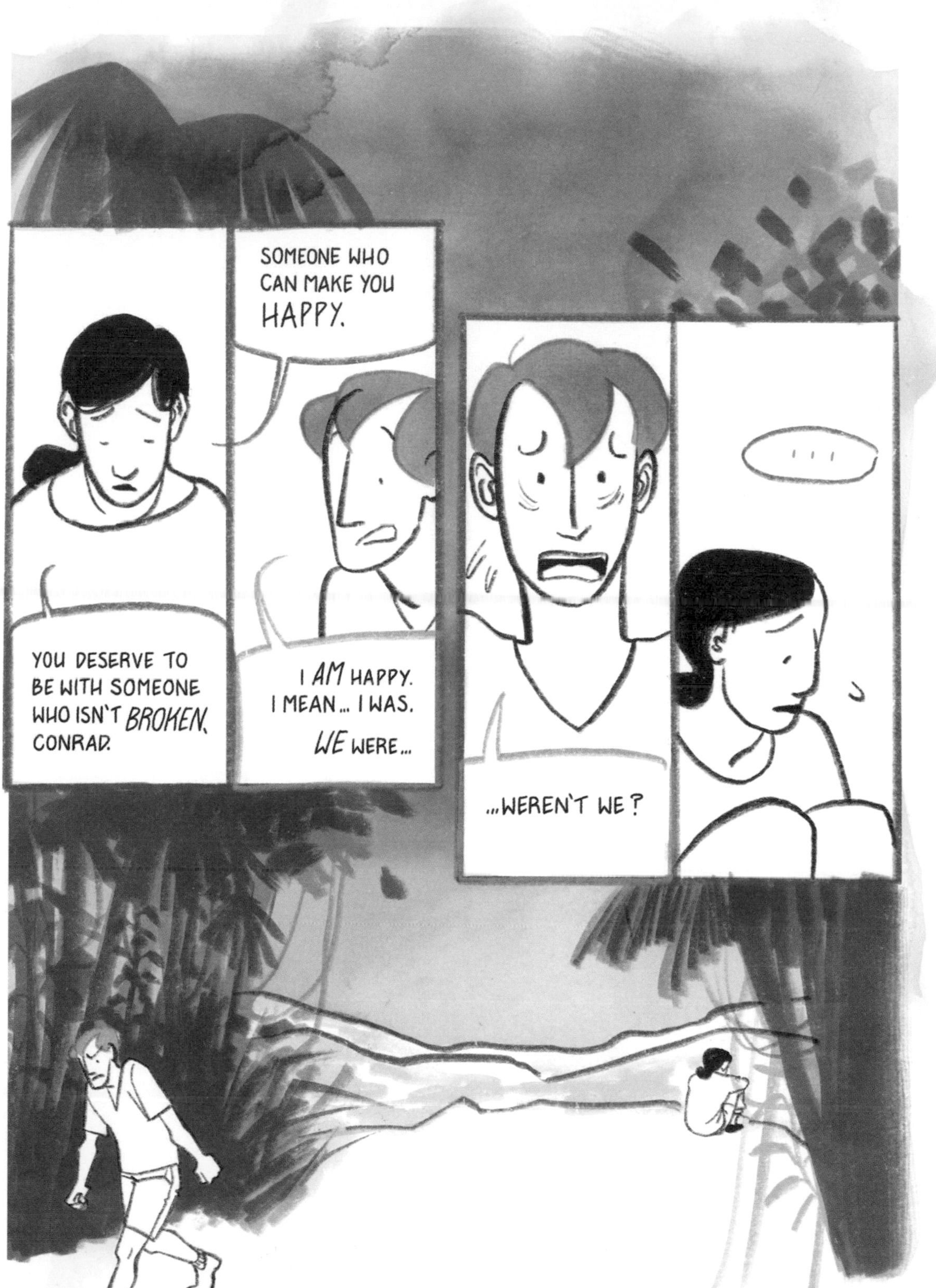
YOU DESERVE TO BE WITH SOMEONE WHO ISN'T *BROKEN*, CONRAD.
SOMEONE WHO CAN MAKE YOU HAPPY.
I *AM* HAPPY. I MEAN... I WAS. *WE* WERE...
...WEREN'T WE?
...

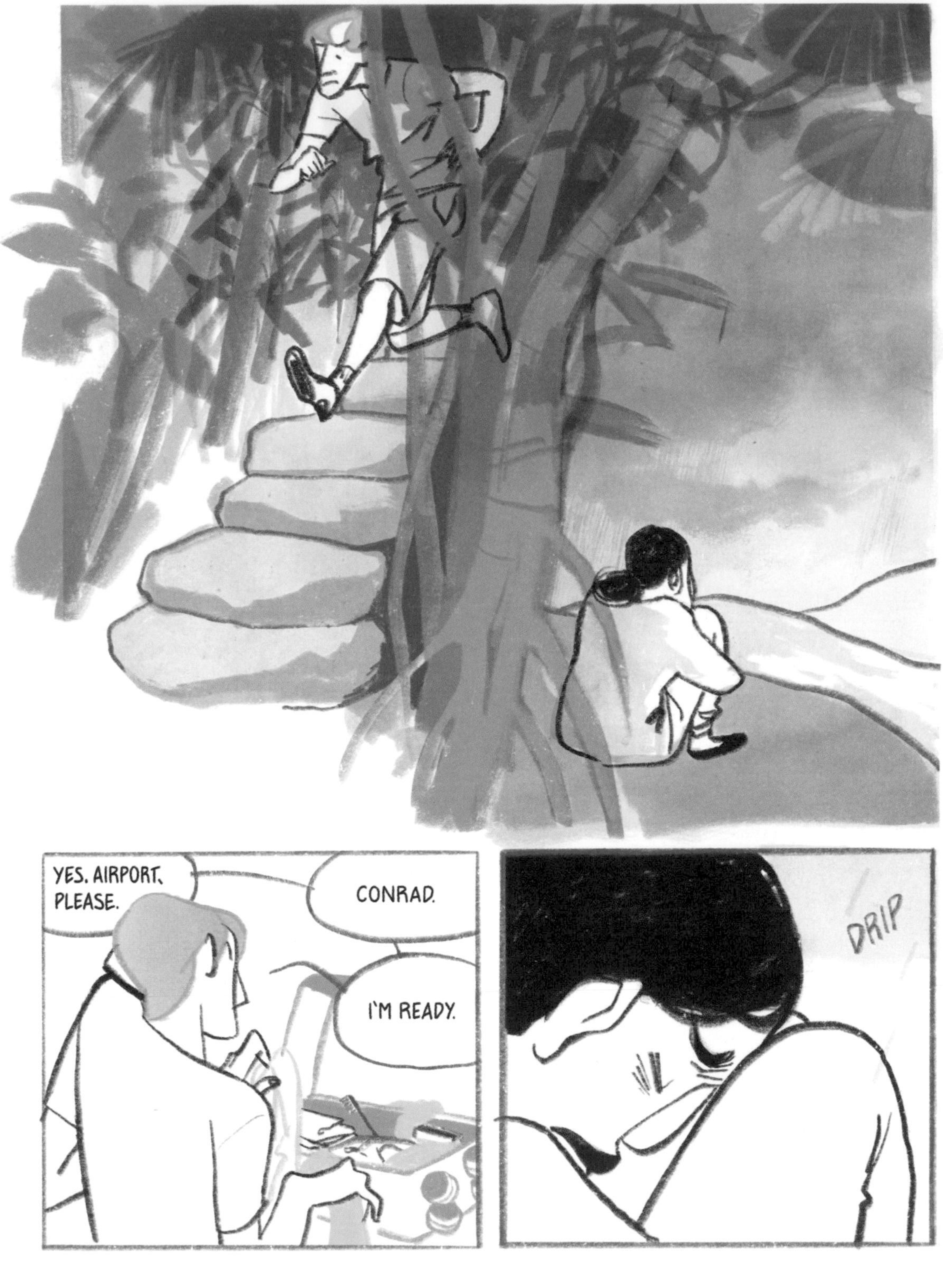
YES. AIRPORT, PLEASE.
CONRAD.
I'M READY.
DRIP

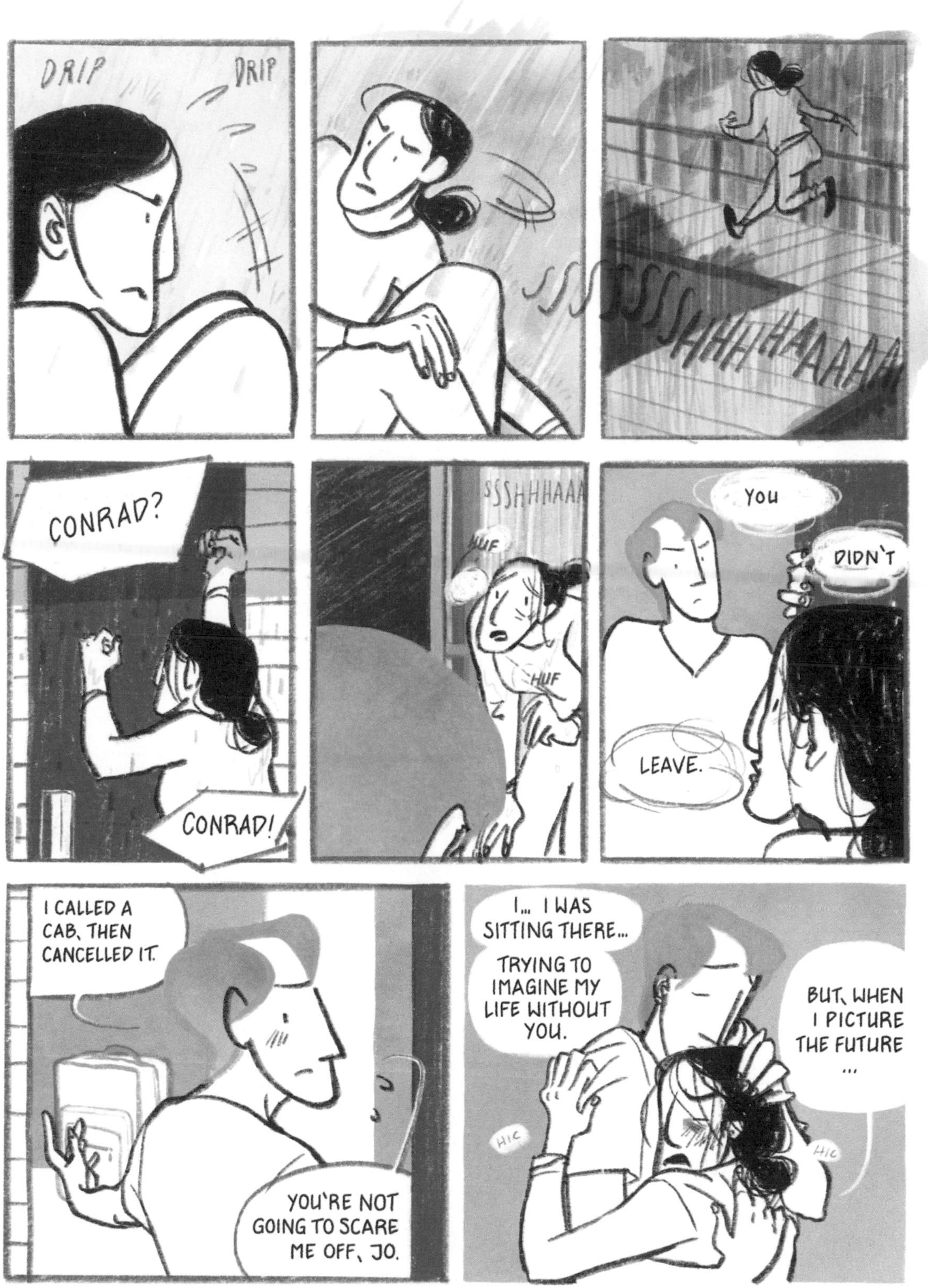
DRIP
DRIP
SSSSSSHHHHAAAAA
CONRAD?
CONRAD!
SSSHHHAAA
HUF
HUF
YOU
DIDN'T
LEAVE.
I CALLED A CAB, THEN CANCELLED IT.
YOU'RE NOT GOING TO SCARE ME OFF, JO.
I... I WAS SITTING THERE...
TRYING TO IMAGINE MY LIFE WITHOUT YOU.
BUT, WHEN I PICTURE THE FUTURE ...
HIC
HIC

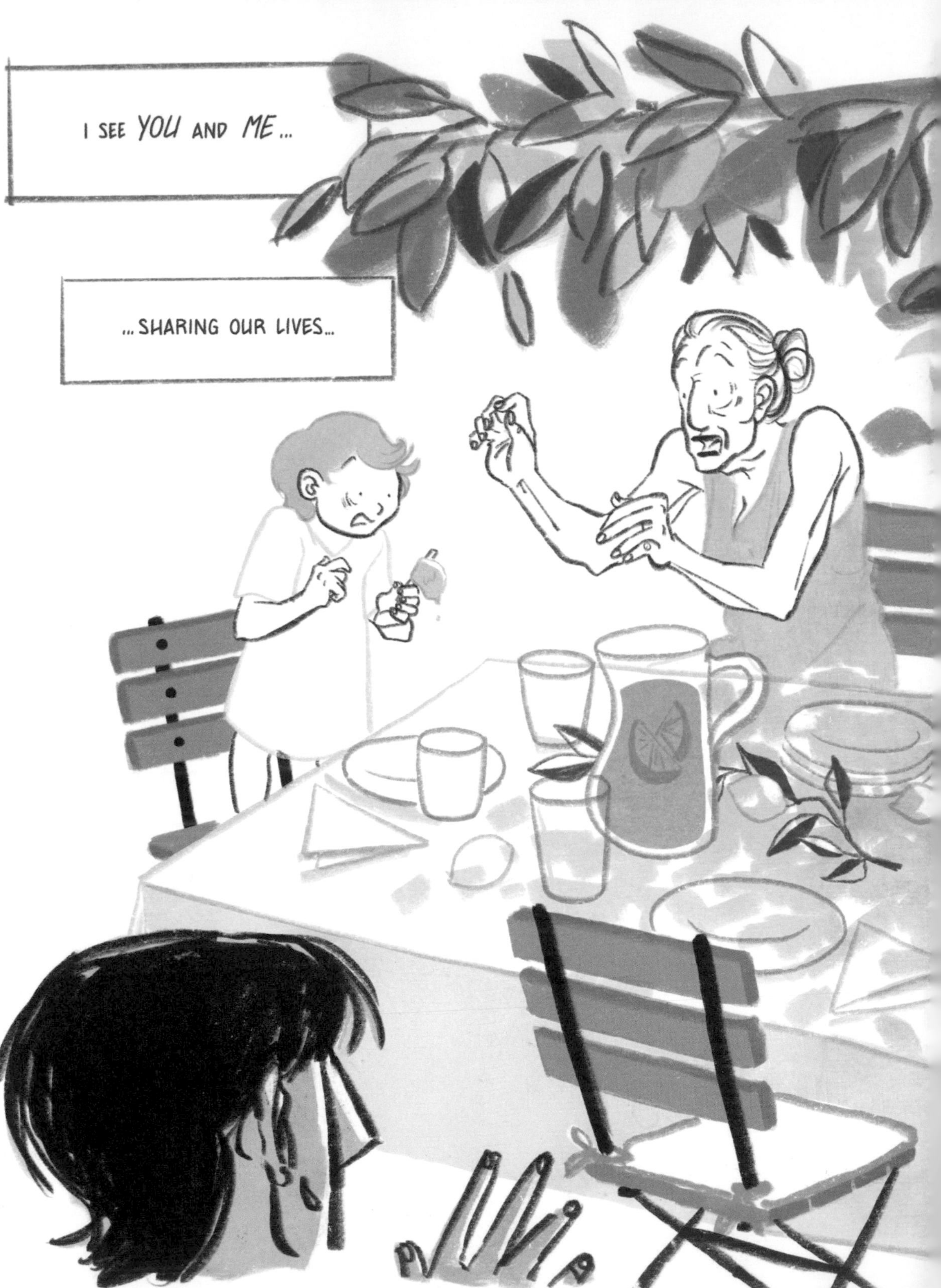
I SEE YOU AND ME...
...SHARING OUR LIVES...

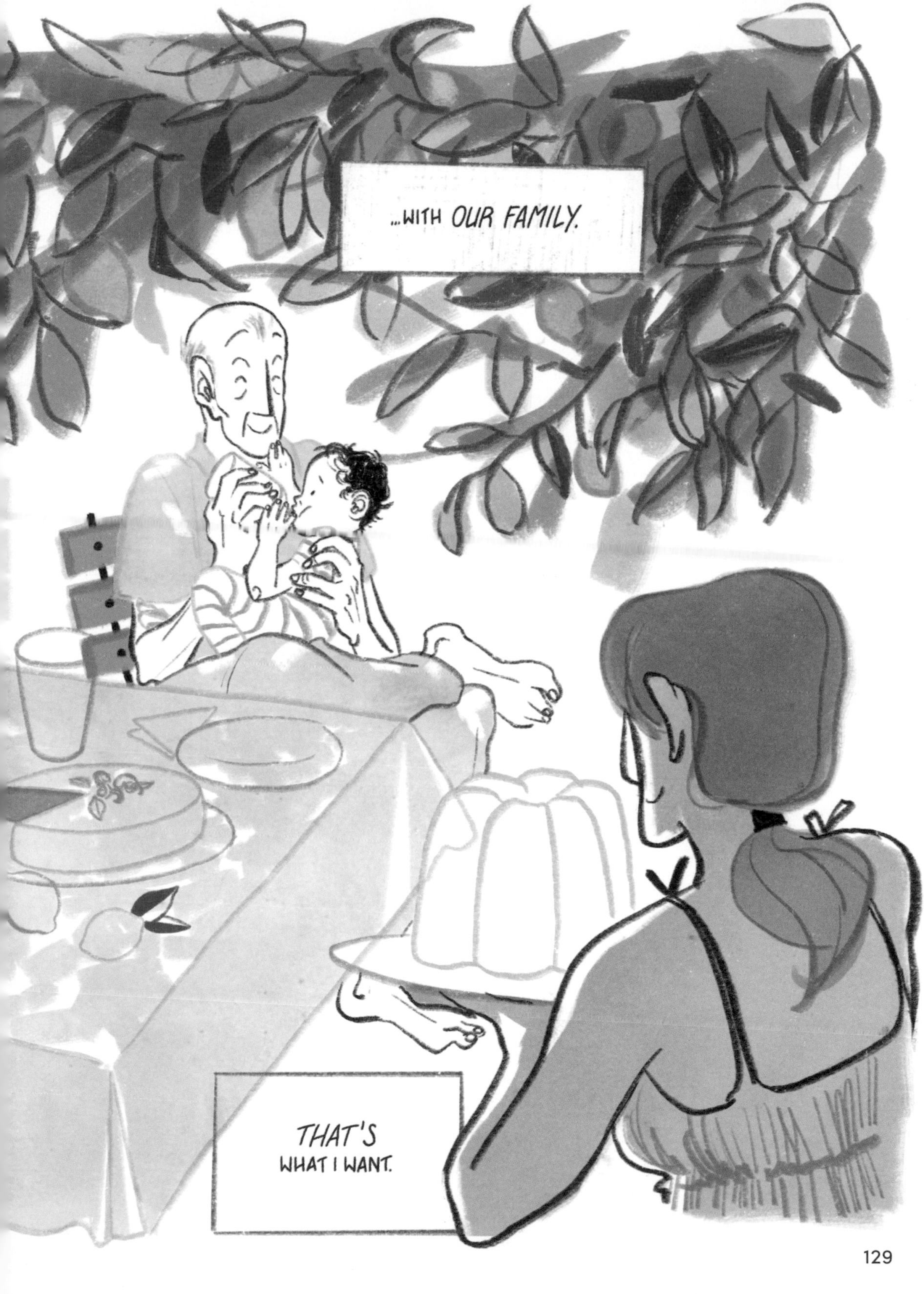
...WITH OUR FAMILY.
THAT'S WHAT I WANT.

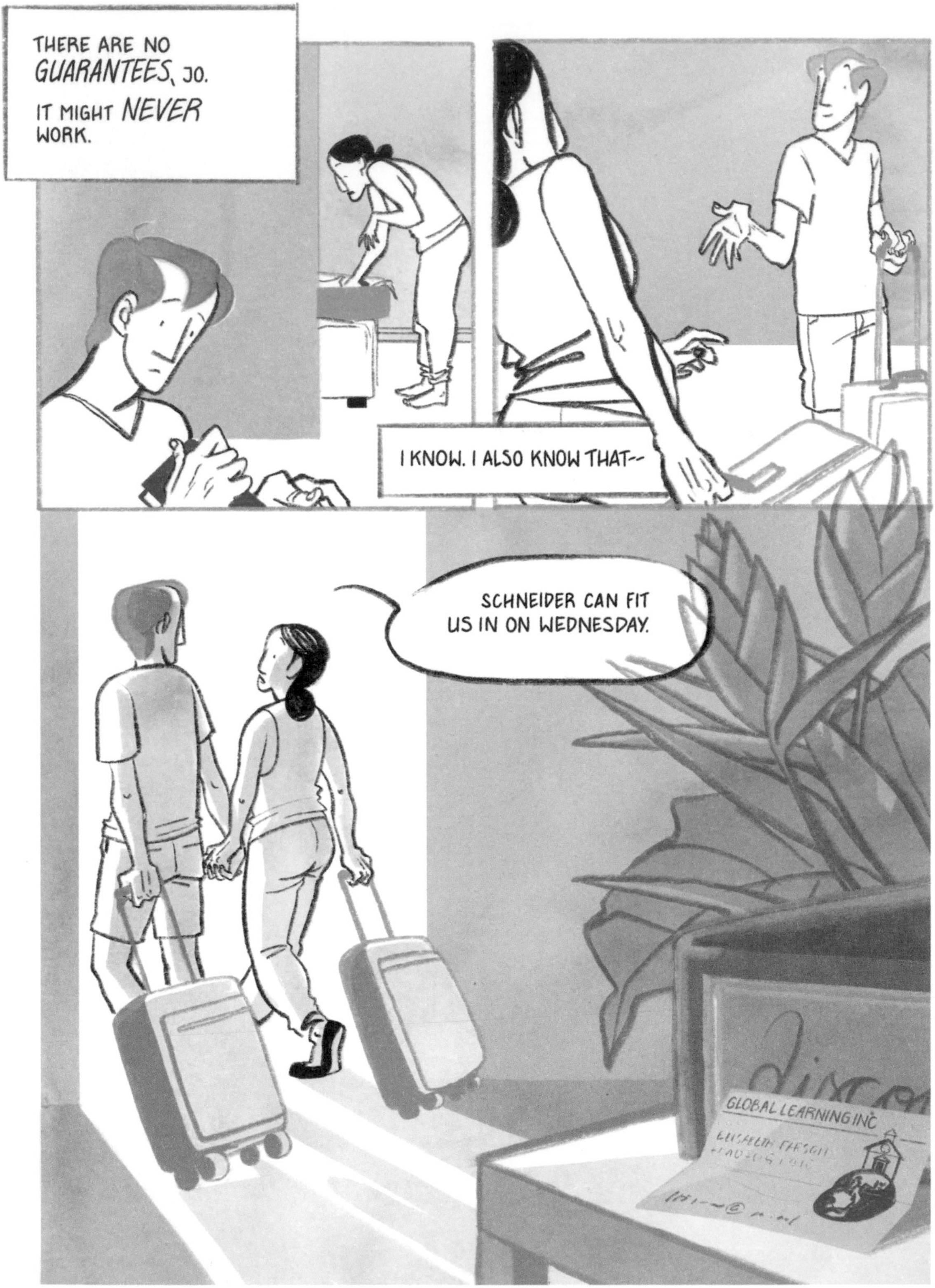
THERE ARE NO GUARANTEES, JO. IT MIGHT NEVER WORK.
I KNOW. I ALSO KNOW THAT--
SCHNEIDER CAN FIT US IN ON WEDNESDAY.
GLOBAL LEARNING INC

CHAPTER 7:

TWO—WEEK WAIT

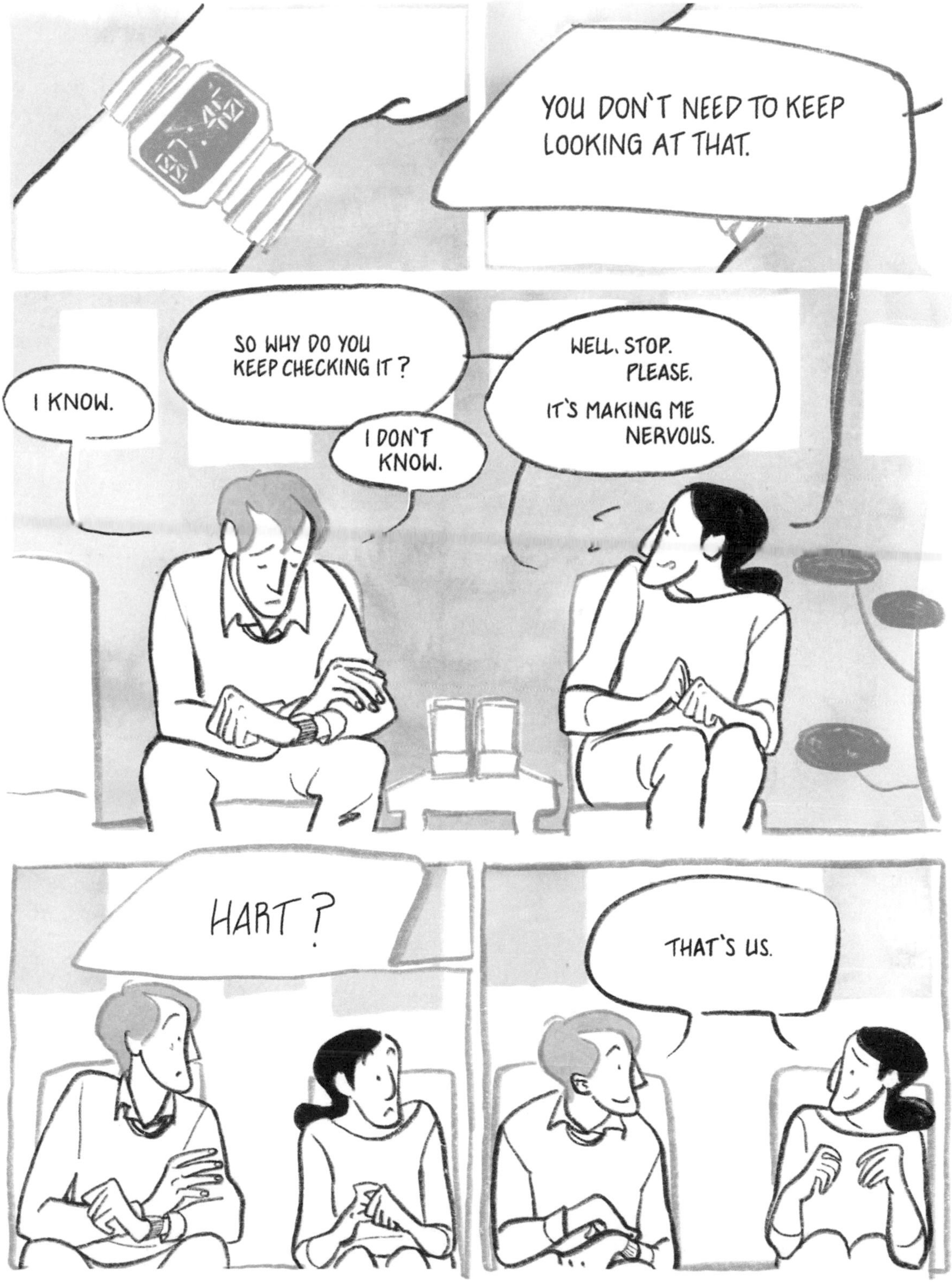
YOU DON'T NEED TO KEEP LOOKING AT THAT.
I KNOW.
SO WHY DO YOU KEEP CHECKING IT ?
I DON'T KNOW.
WELL. STOP. PLEASE.
IT'S MAKING ME NERVOUS.
HART ?
THAT'S US.

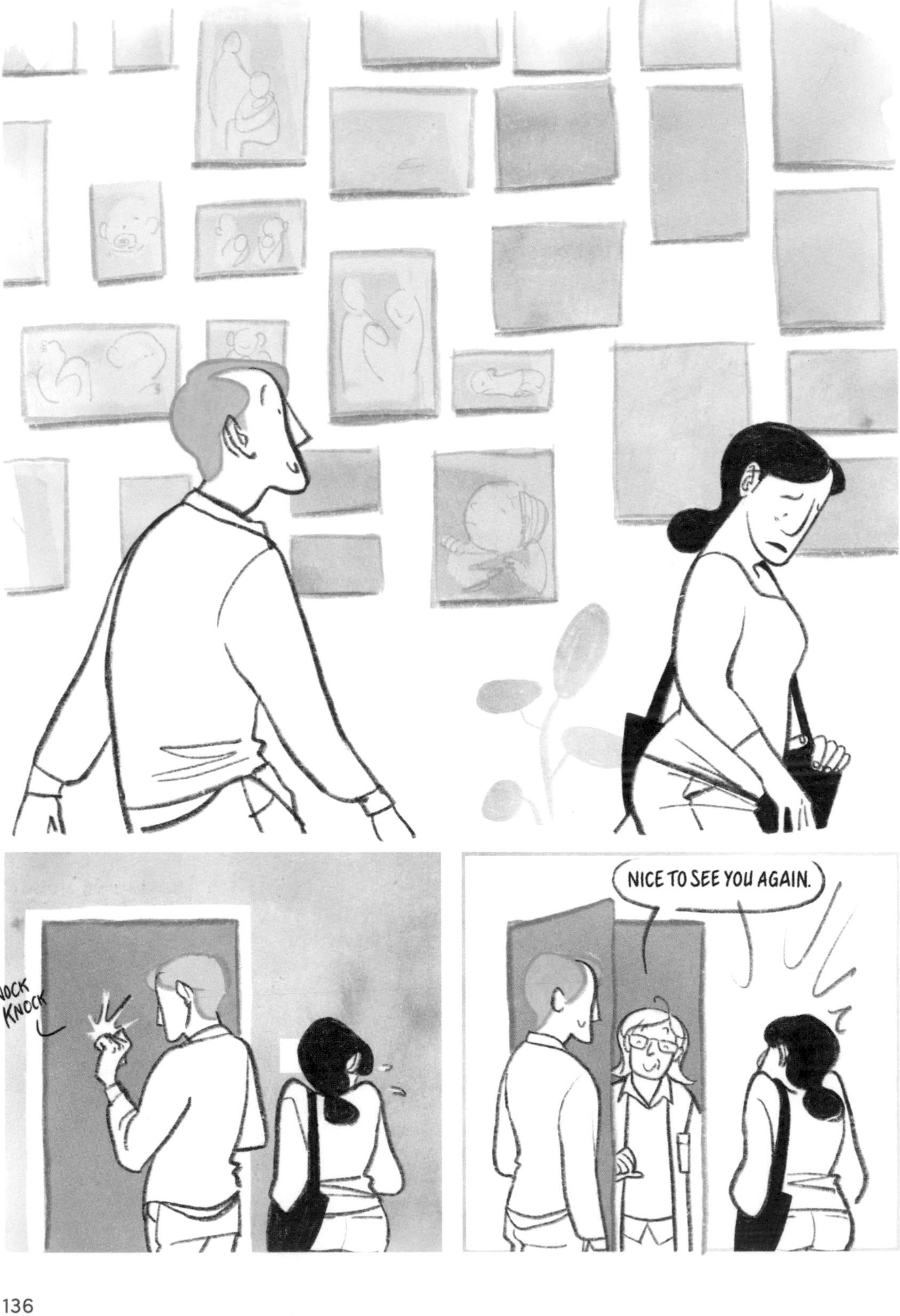
KNOCK KNOCK
NICE TO SEE YOU AGAIN.

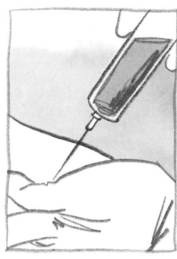

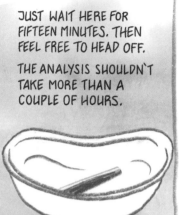

JUST WAIT HERE FOR FIFTEEN MINUTES. THEN FEEL FREE TO HEAD OFF.

THE ANALYSIS SHOULDN'T TAKE MORE THAN A COUPLE OF HOURS.

THANKS, I'LL BE ON MY MOBILE.

GOOD—

—LUCK!

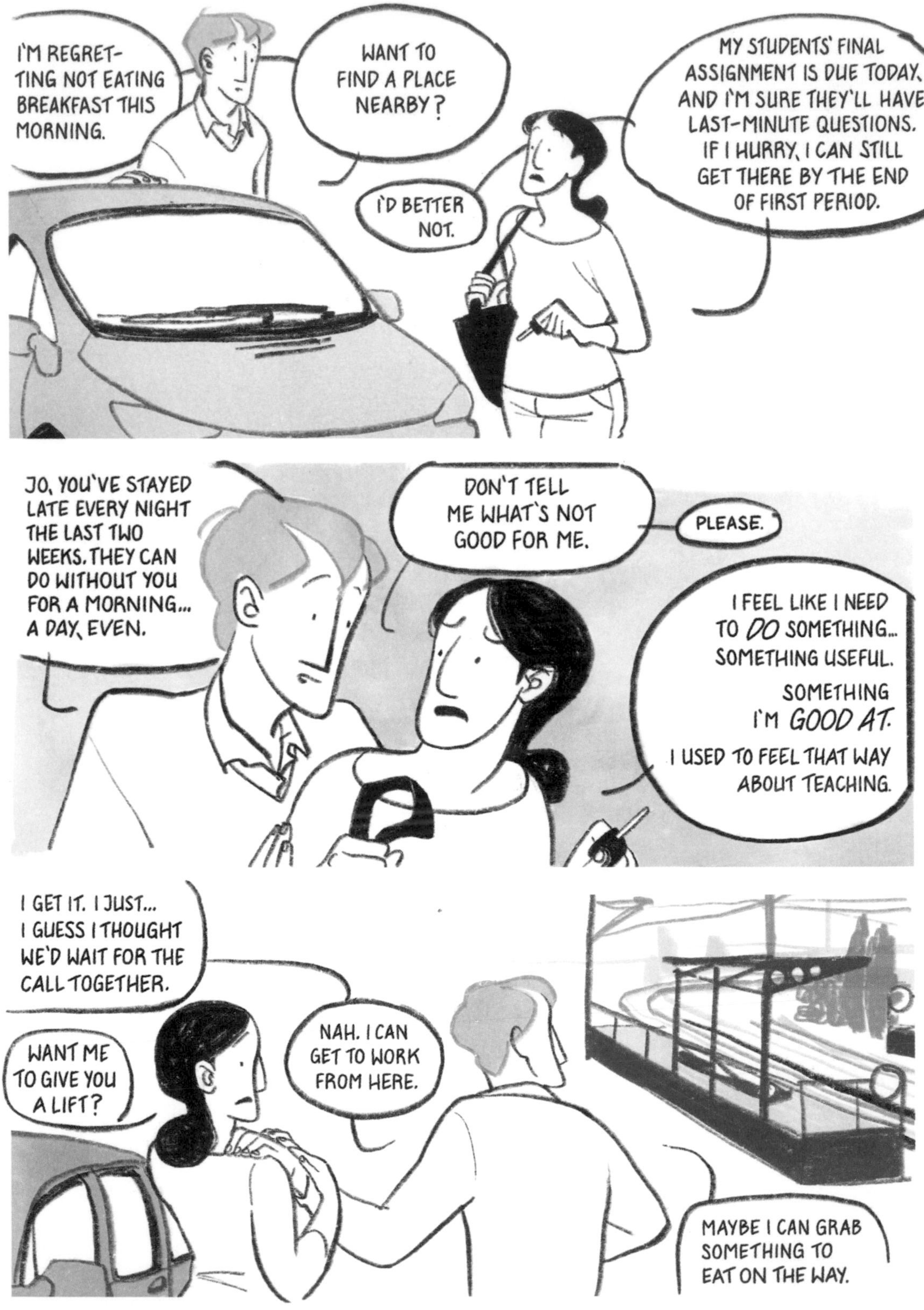
I'M REGRETTING NOT EATING BREAKFAST THIS MORNING.
WANT TO FIND A PLACE NEARBY?
I'D BETTER NOT.
MY STUDENTS' FINAL ASSIGNMENT IS DUE TODAY, AND I'M SURE THEY'LL HAVE LAST-MINUTE QUESTIONS. IF I HURRY, I CAN STILL GET THERE BY THE END OF FIRST PERIOD.
JO, YOU'VE STAYED LATE EVERY NIGHT THE LAST TWO WEEKS. THEY CAN DO WITHOUT YOU FOR A MORNING... A DAY, EVEN.
DON'T TELL ME WHAT'S NOT GOOD FOR ME.
PLEASE.
I FEEL LIKE I NEED TO *DO* SOMETHING... SOMETHING USEFUL.
SOMETHING I'M *GOOD AT.*
I USED TO FEEL THAT WAY ABOUT TEACHING.
I GET IT. I JUST... I GUESS I THOUGHT WE'D WAIT FOR THE CALL TOGETHER.
WANT ME TO GIVE YOU A LIFT?
NAH. I CAN GET TO WORK FROM HERE.
MAYBE I CAN GRAB SOMETHING TO EAT ON THE WAY.

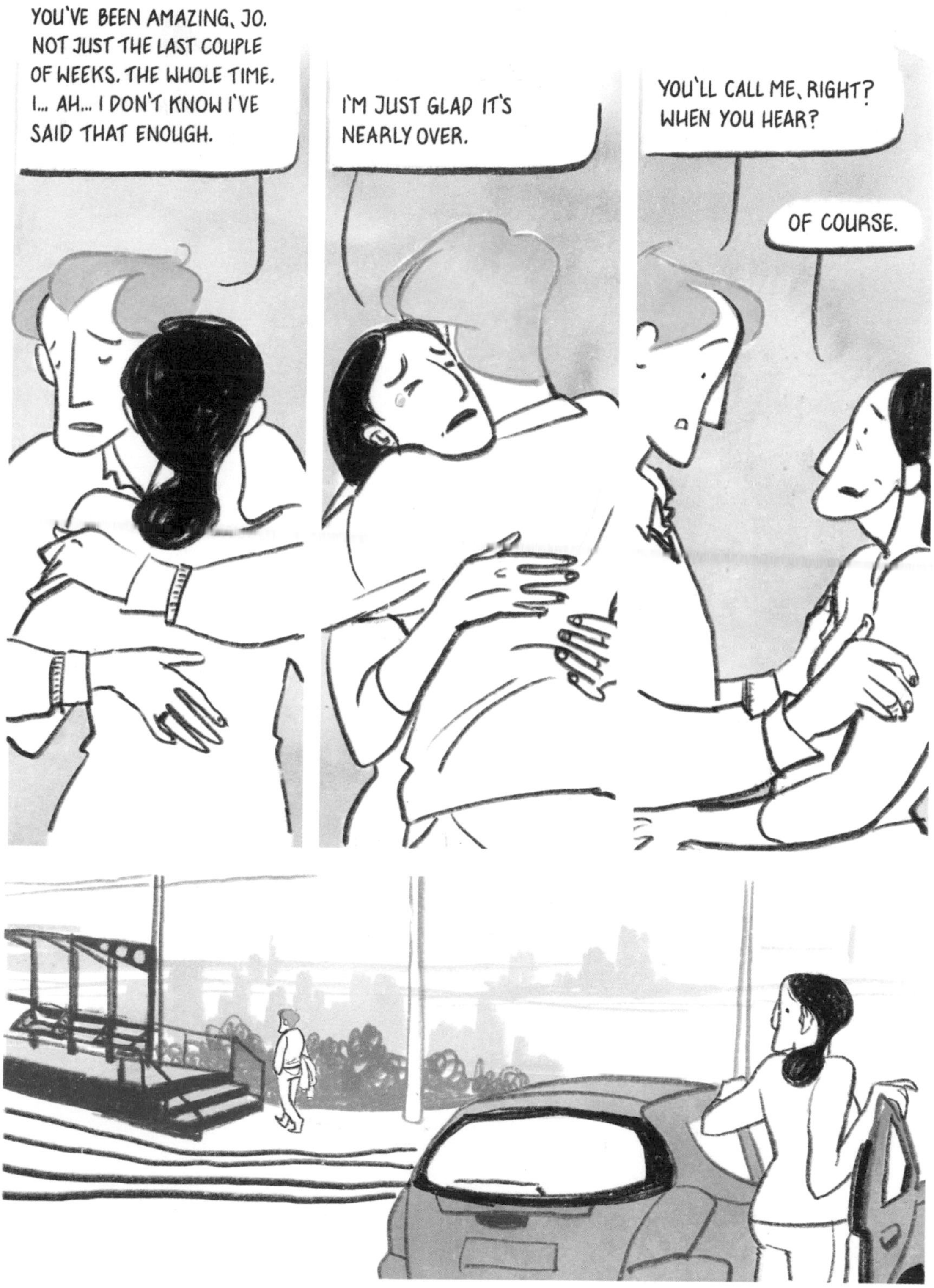
YOU'VE BEEN AMAZING, JO. NOT JUST THE LAST COUPLE OF WEEKS. THE WHOLE TIME. I... AH... I DON'T KNOW I'VE SAID THAT ENOUGH.
I'M JUST GLAD IT'S NEARLY OVER.
YOU'LL CALL ME, RIGHT? WHEN YOU HEAR?
OF COURSE.

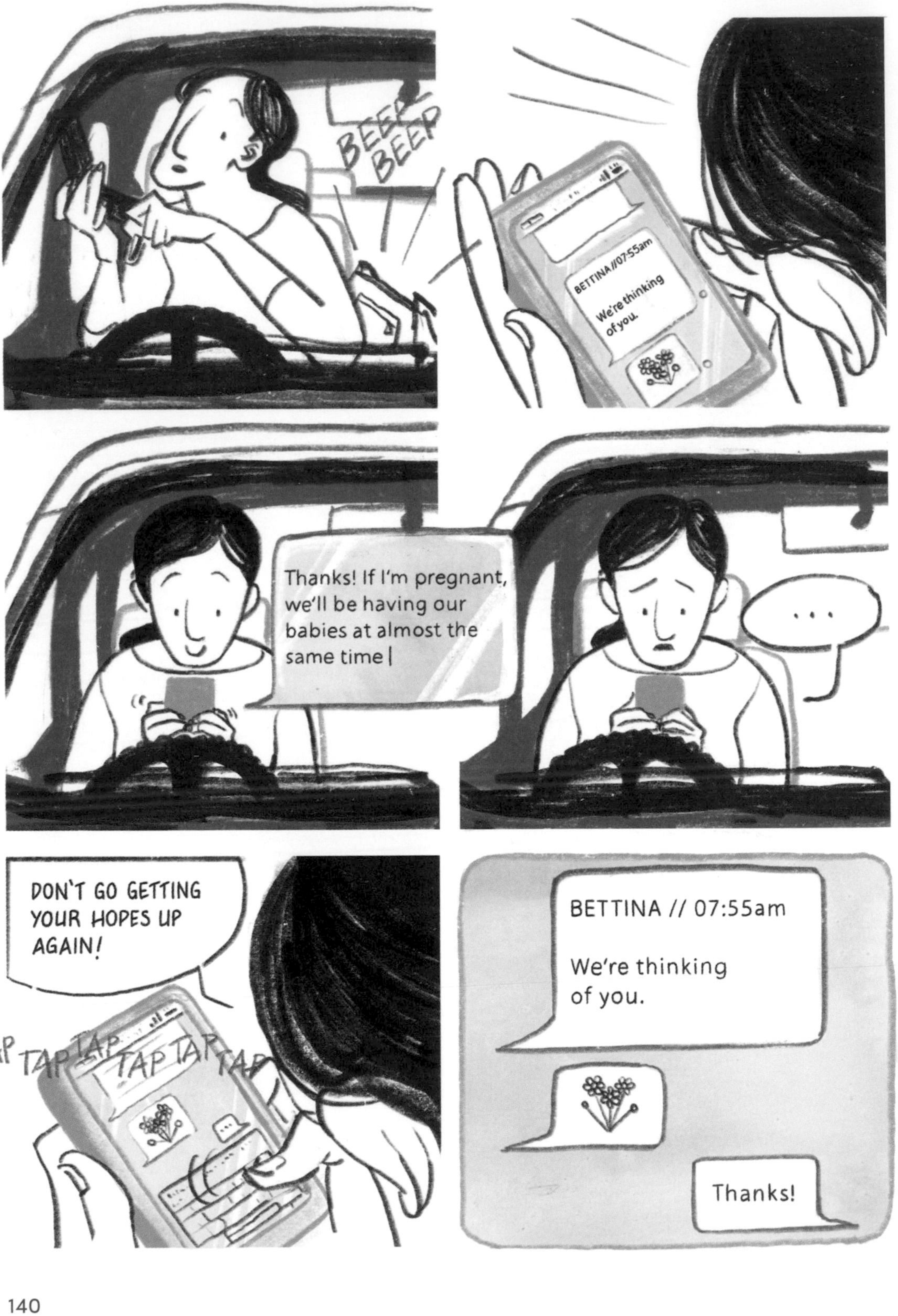
BEEP BEEP
BETTINA //07:55am
We're thinking of you.
Thanks! If I'm pregnant, we'll be having our babies at almost the same time
. . .
DON'T GO GETTING YOUR HOPES UP AGAIN!
TAP TAP TAP TAP TAP TAP
BETTINA // 07:55am
We're thinking of you.
Thanks!

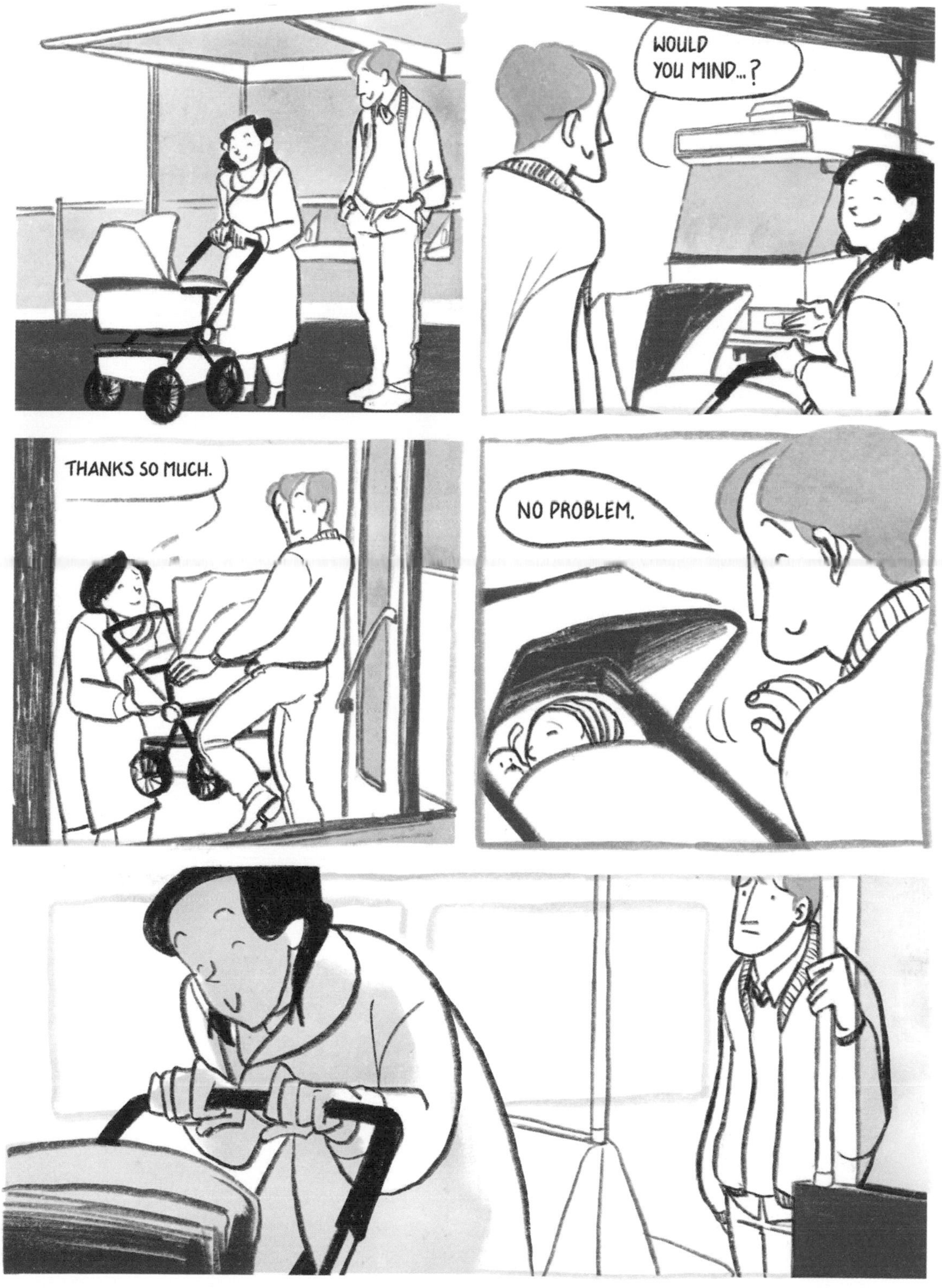
WOULD YOU MIND...?
THANKS SO MUCH.
NO PROBLEM.

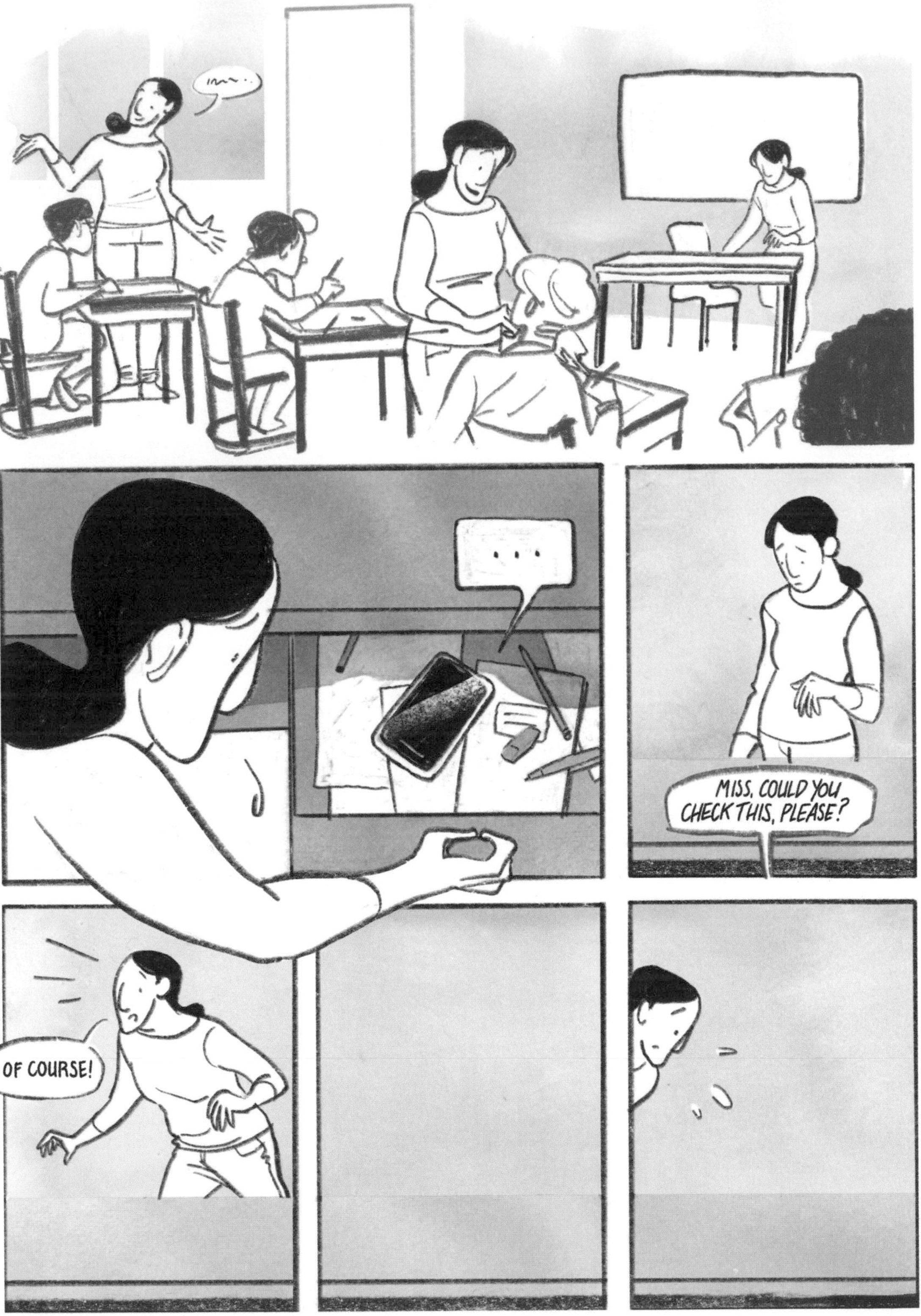
MISS, COULD YOU, CHECK THIS, PLEASE?
OF COURSE!

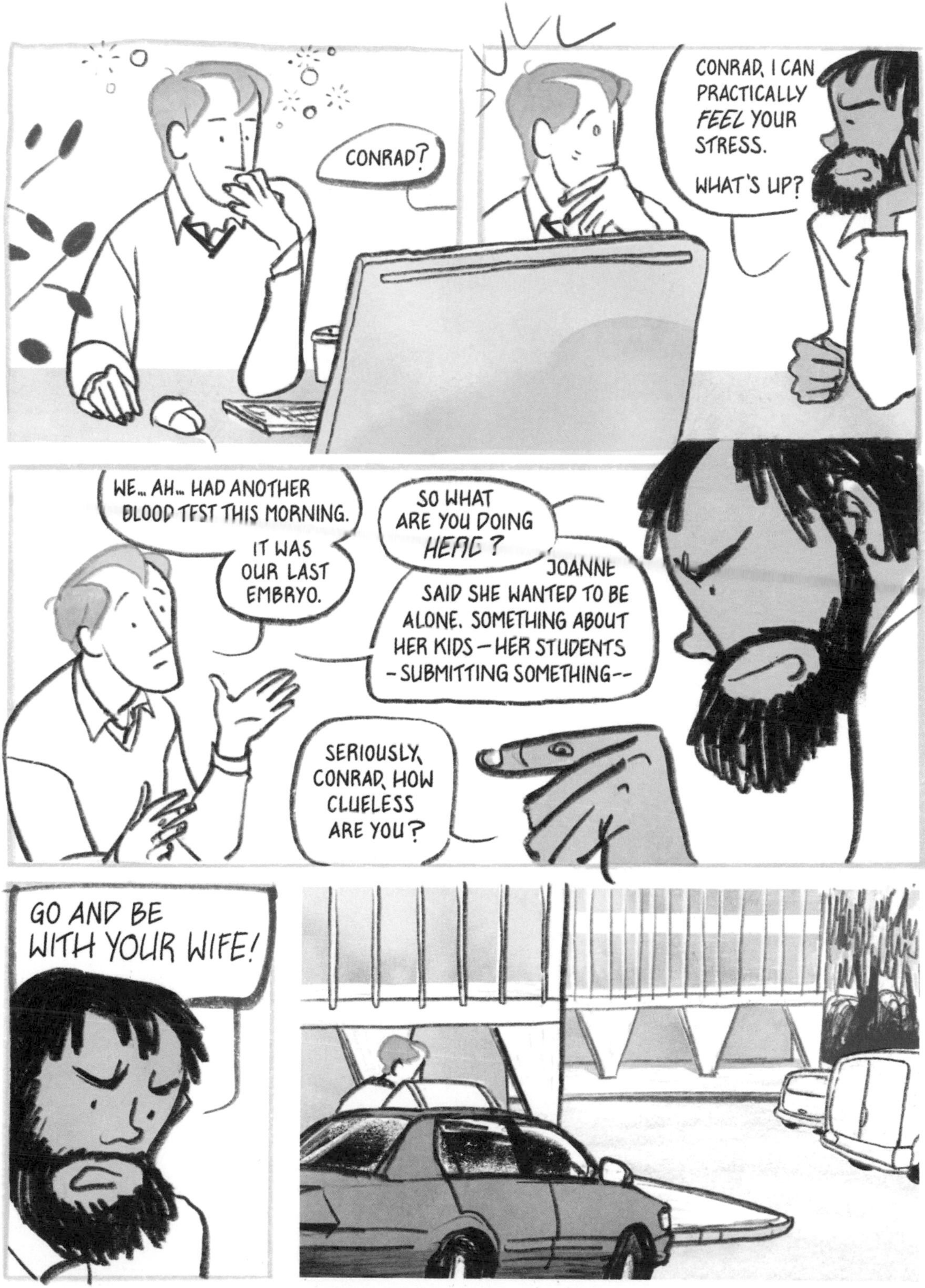
CONRAD?
CONRAD, I CAN PRACTICALLY FEEL YOUR STRESS.
WHAT'S UP?
WE... AH... HAD ANOTHER BLOOD TEST THIS MORNING.
IT WAS OUR LAST EMBRYO.
SO WHAT ARE YOU DOING HERE?
JOANNE SAID SHE WANTED TO BE ALONE. SOMETHING ABOUT HER KIDS—HER STUDENTS—SUBMITTING SOMETHING--
SERIOUSLY, CONRAD, HOW CLUELESS ARE YOU?
GO AND BE WITH YOUR WIFE!

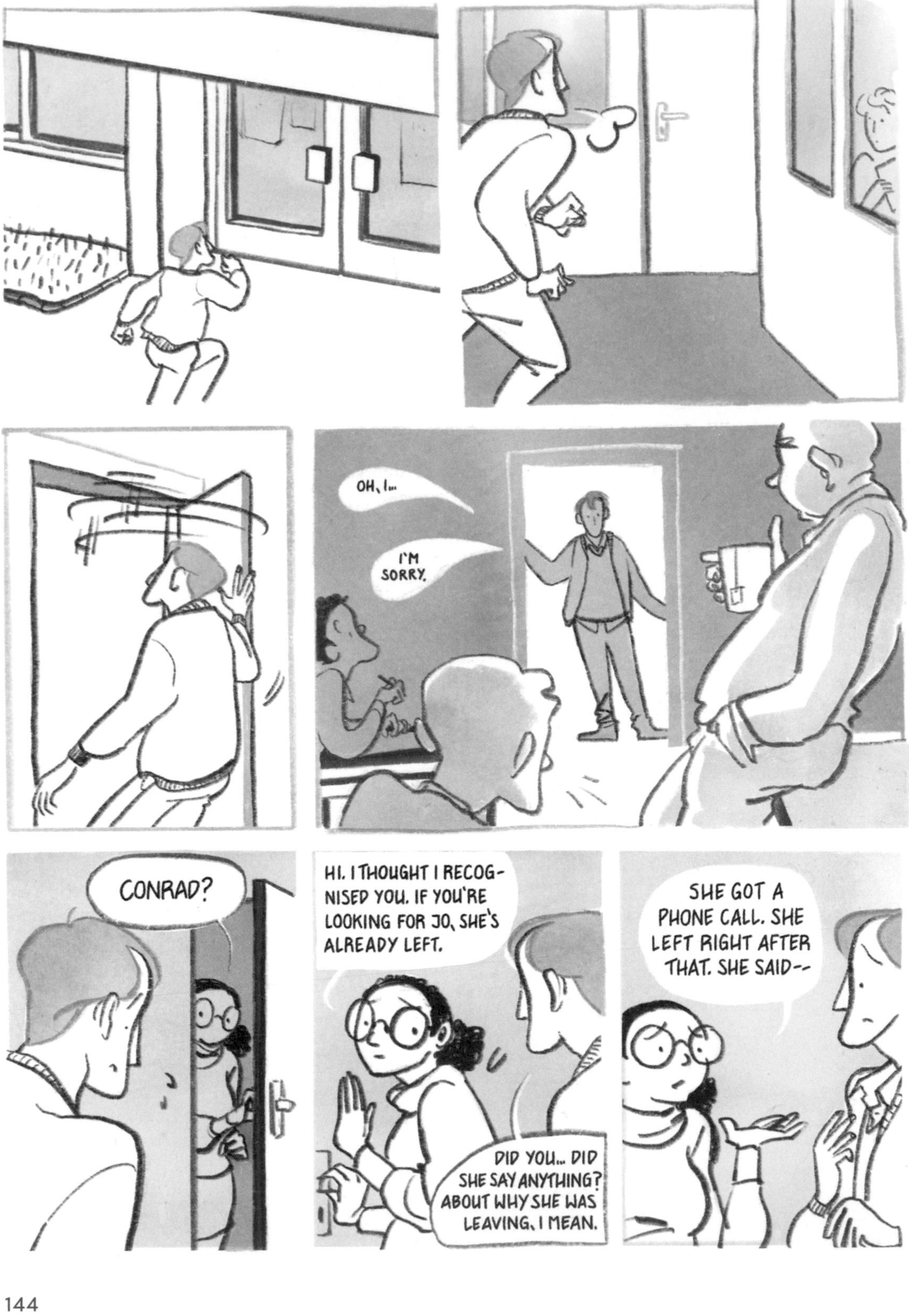

OH, I...
I'M SORRY.
CONRAD?
HI. I THOUGHT I RECOGNISED YOU. IF YOU'RE LOOKING FOR JO, SHE'S ALREADY LEFT.
DID YOU... DID SHE SAY ANYTHING? ABOUT WHY SHE WAS LEAVING, I MEAN.
SHE GOT A PHONE CALL. SHE LEFT RIGHT AFTER THAT. SHE SAID--

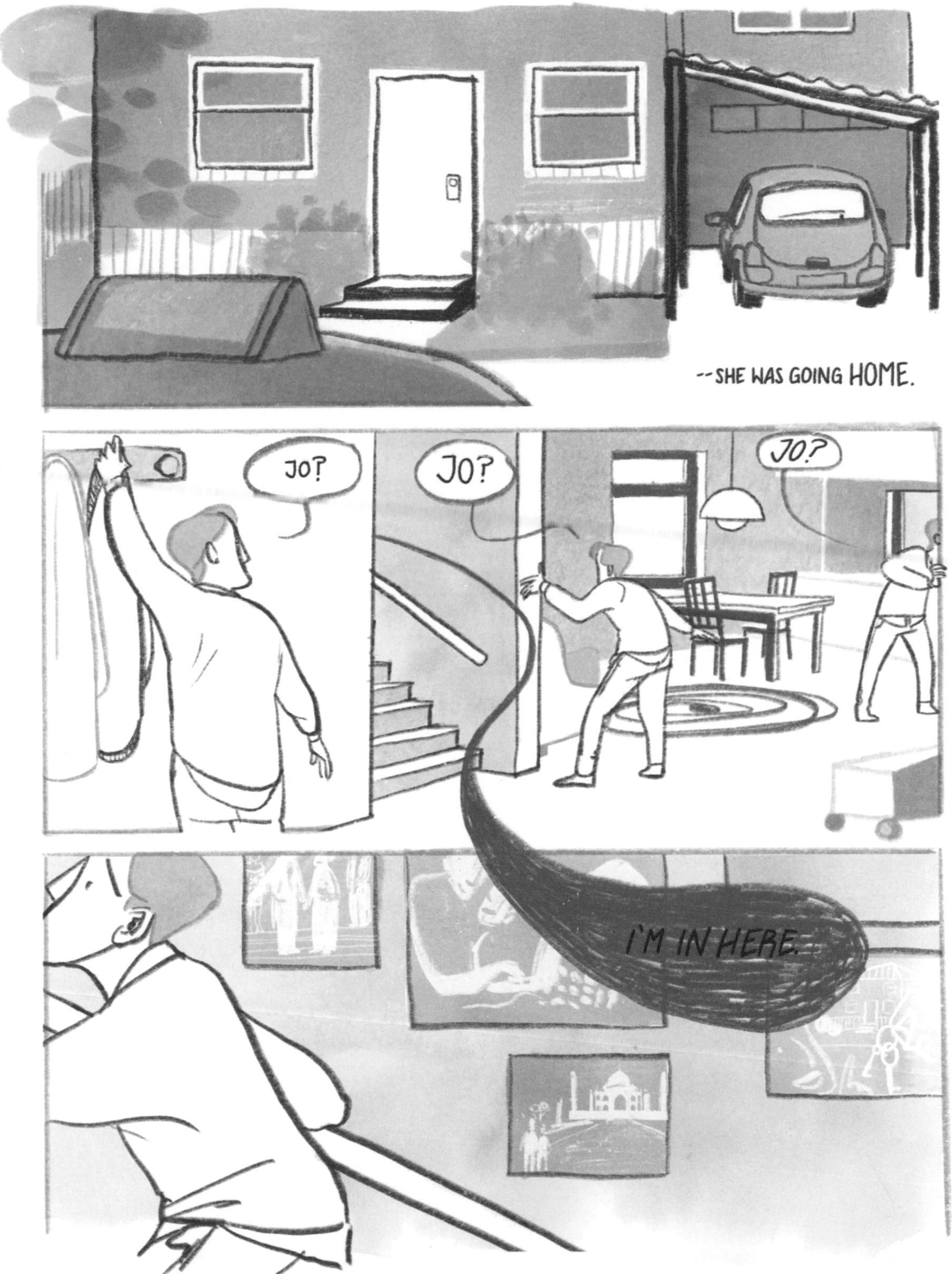
--SHE WAS GOING HOME.
JO?
JO?
JO?
I'M IN HERE.

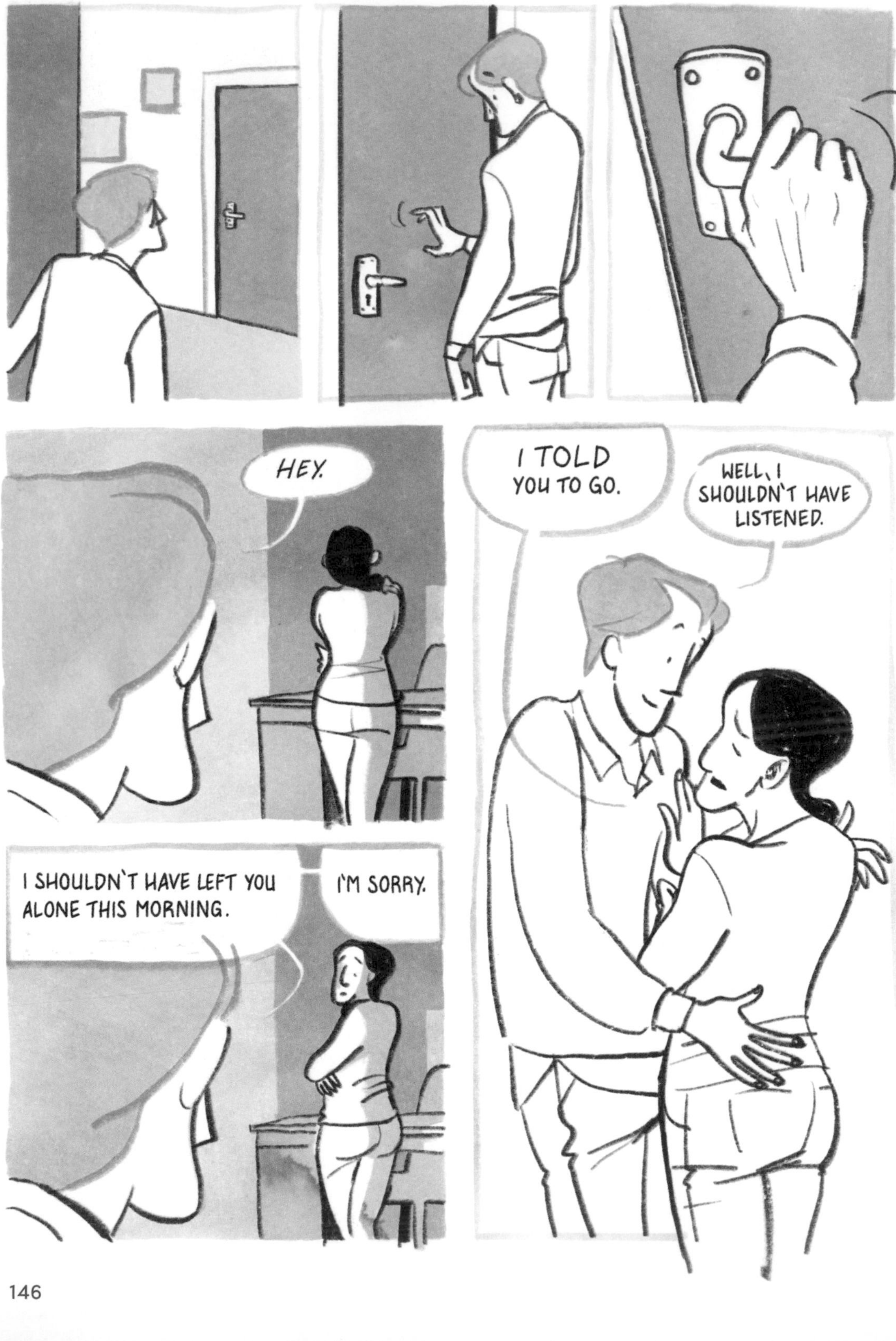
HEY.
I TOLD YOU TO GO.
WELL, I SHOULDN'T HAVE LISTENED.
I SHOULDN'T HAVE LEFT YOU ALONE THIS MORNING.
I'M SORRY.

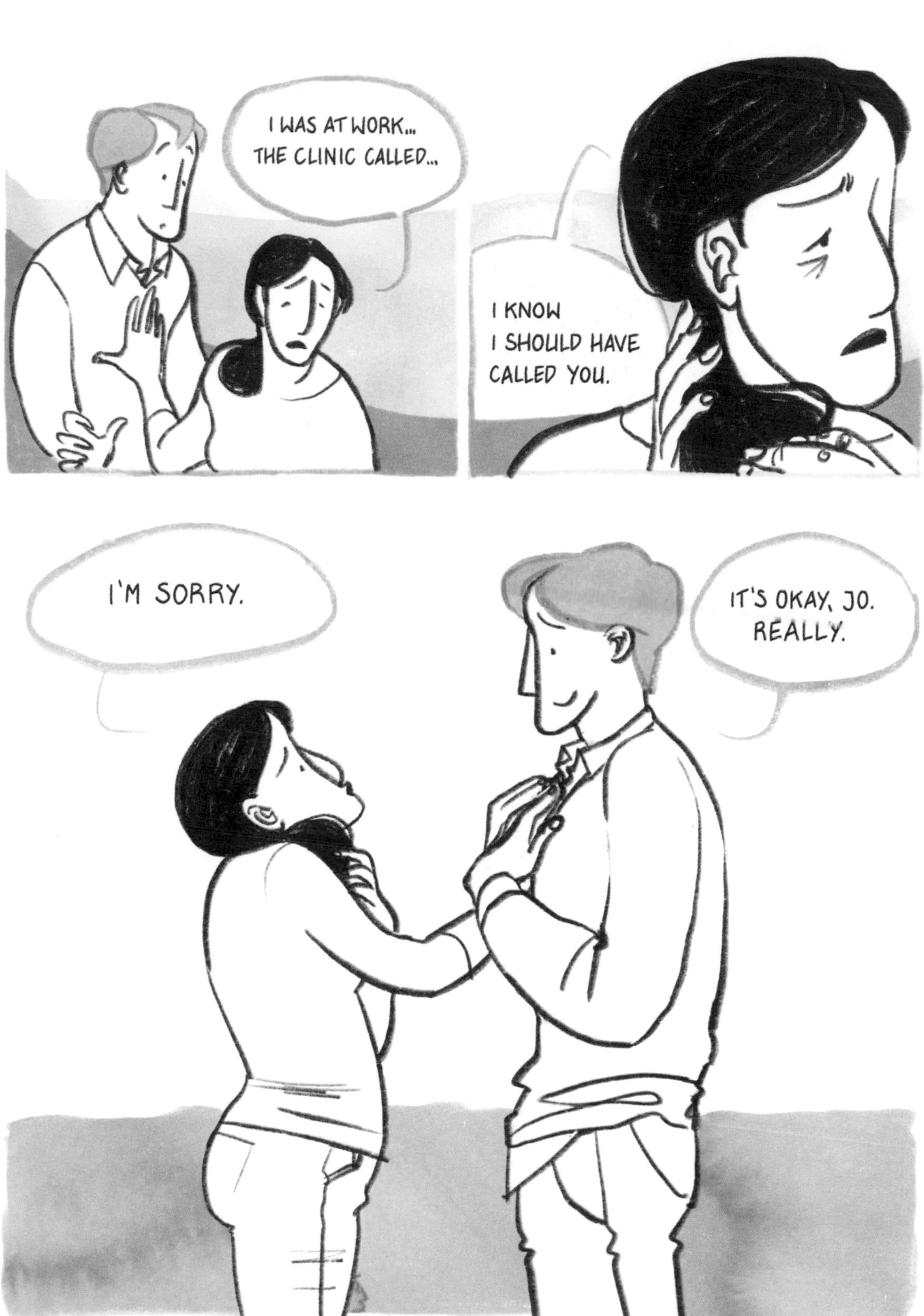
I WAS AT WORK...
THE CLINIC CALLED...
I KNOW
I SHOULD HAVE
CALLED YOU.
I'M SORRY.
IT'S OKAY, JO.
REALLY.

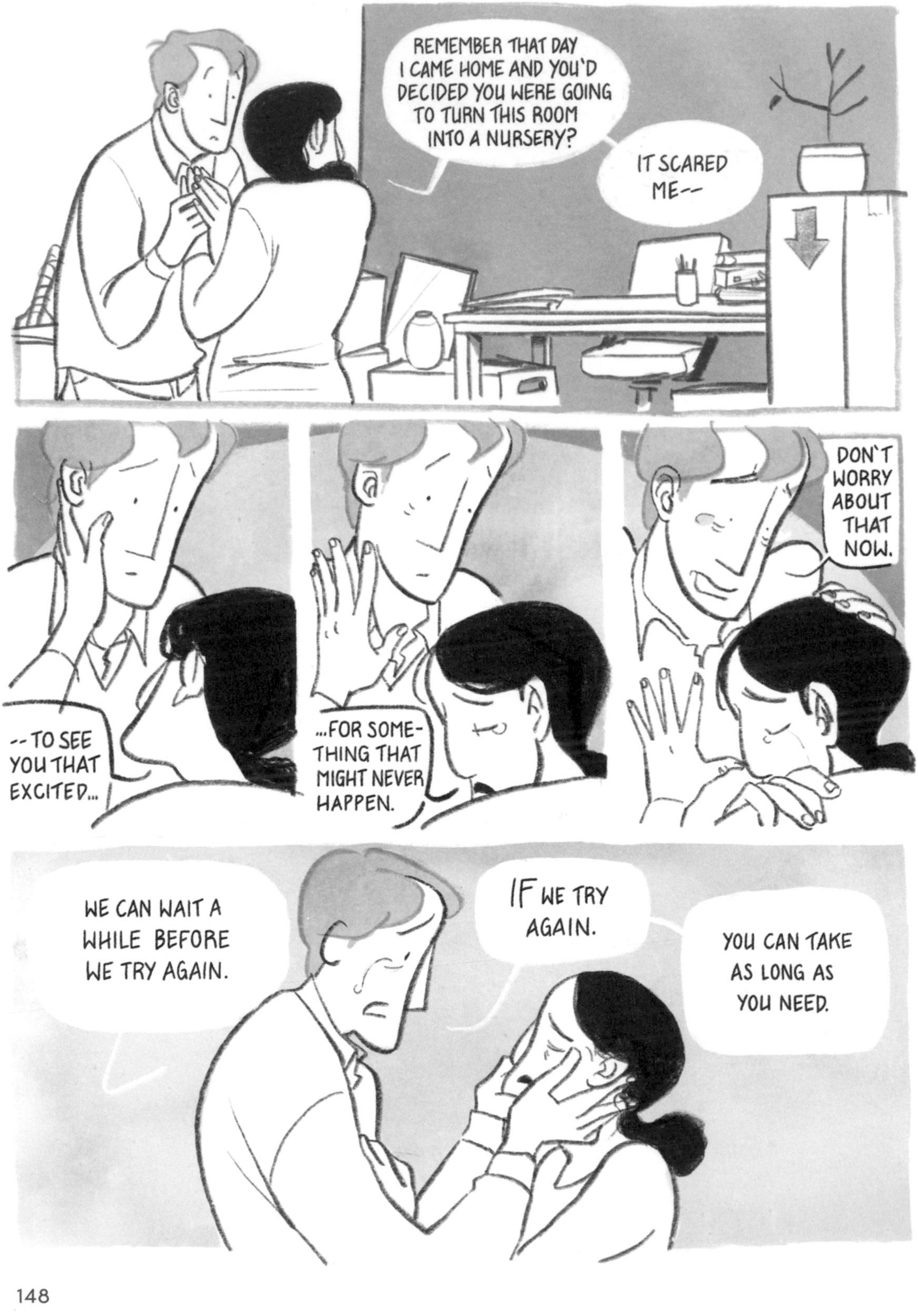
REMEMBER THAT DAY I CAME HOME AND YOU'D DECIDED YOU WERE GOING TO TURN THIS ROOM INTO A NURSERY?
IT SCARED ME--
--TO SEE YOU THAT EXCITED...
...FOR SOMETHING THAT MIGHT NEVER HAPPEN.
DON'T WORRY ABOUT THAT NOW.
WE CAN WAIT A WHILE BEFORE WE TRY AGAIN.
IF WE TRY AGAIN.
YOU CAN TAKE AS LONG AS YOU NEED.

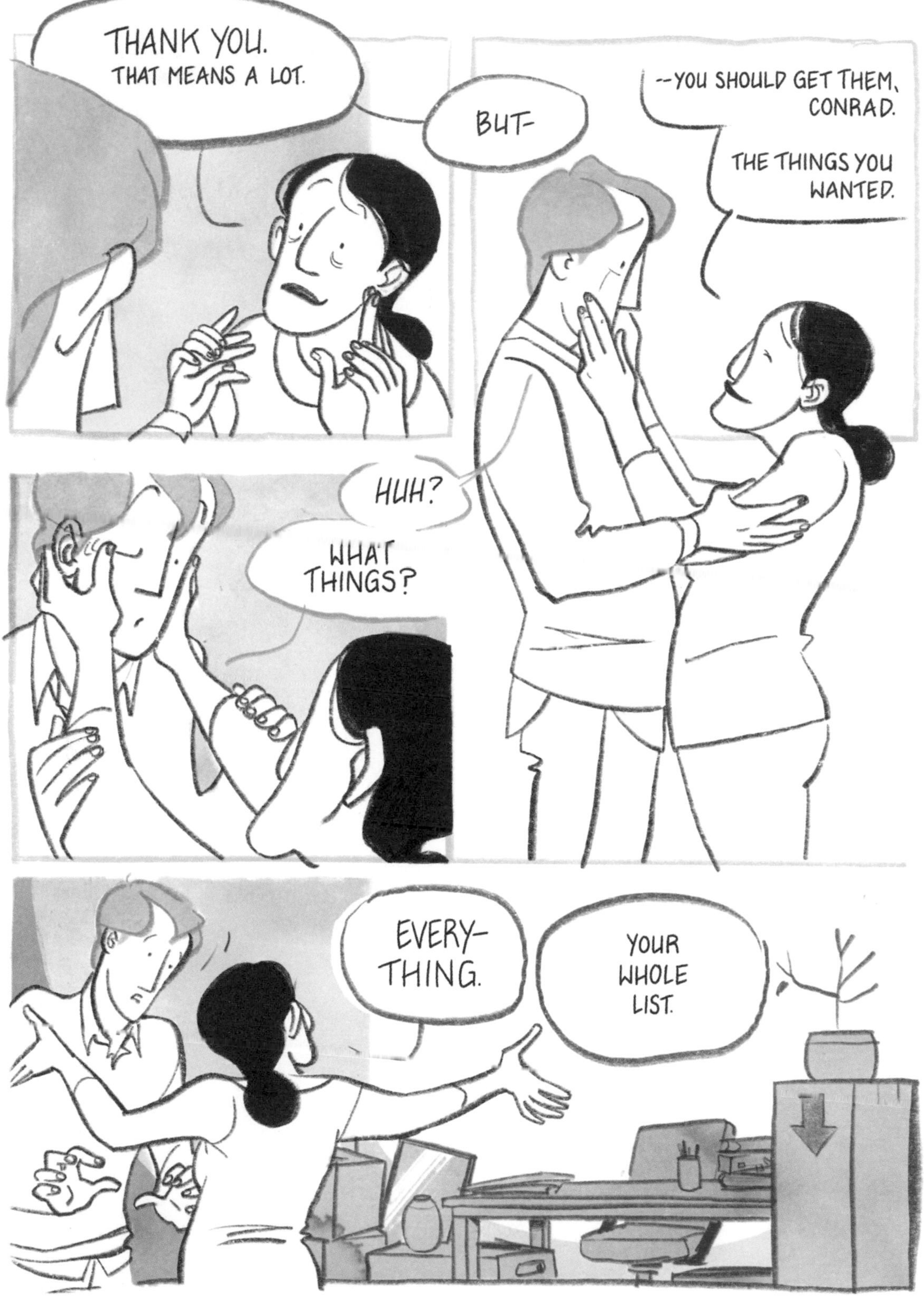
THANK YOU.
THAT MEANS A LOT.
BUT-
--YOU SHOULD GET THEM, CONRAD.
THE THINGS YOU WANTED.
HUH?
WHAT THINGS?
EVERY-THING.
YOUR WHOLE LIST.

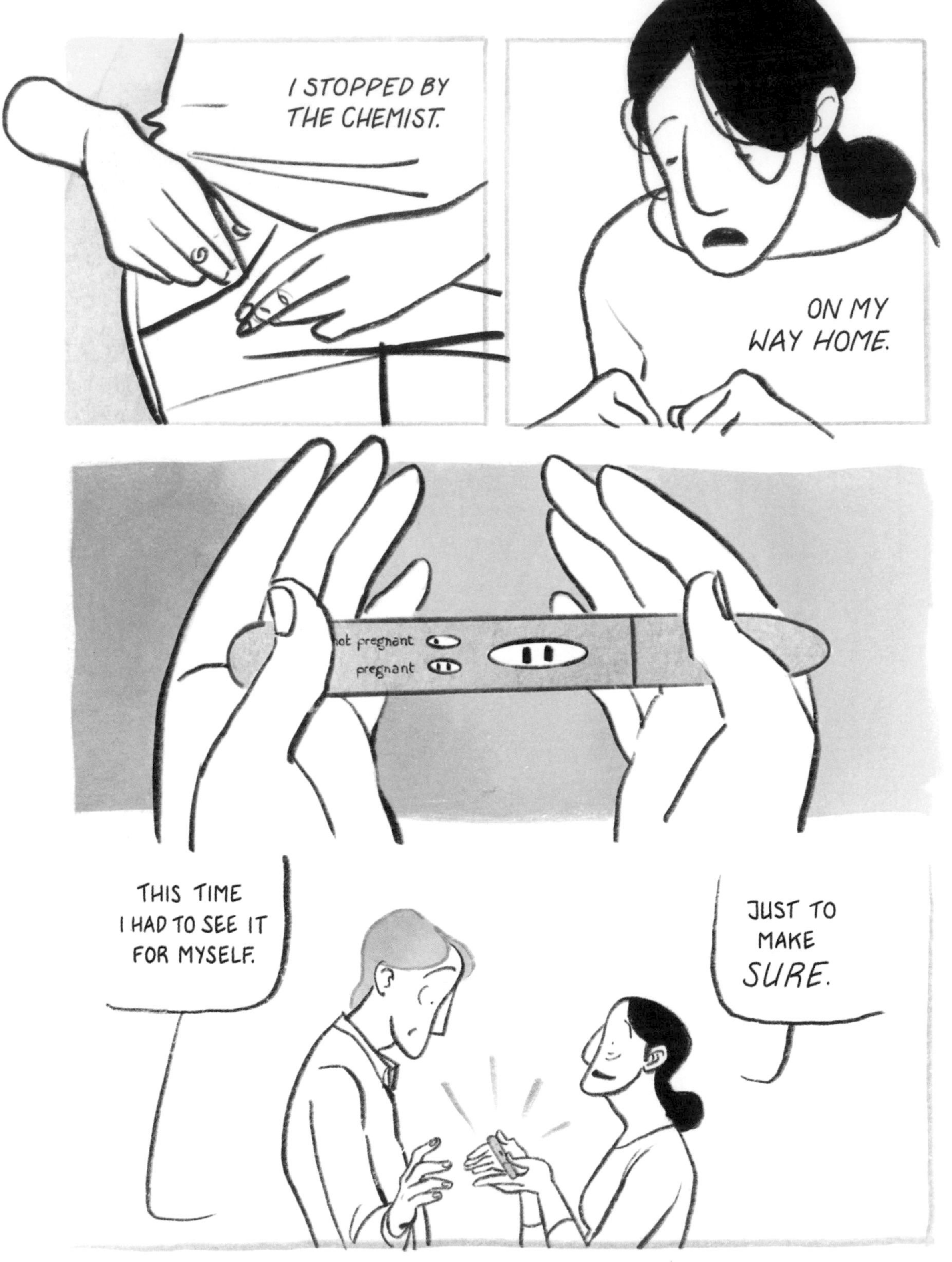
I STOPPED BY THE CHEMIST.
ON MY WAY HOME.
not pregnant
pregnant
THIS TIME I HAD TO SEE IT FOR MYSELF.
JUST TO MAKE SURE.

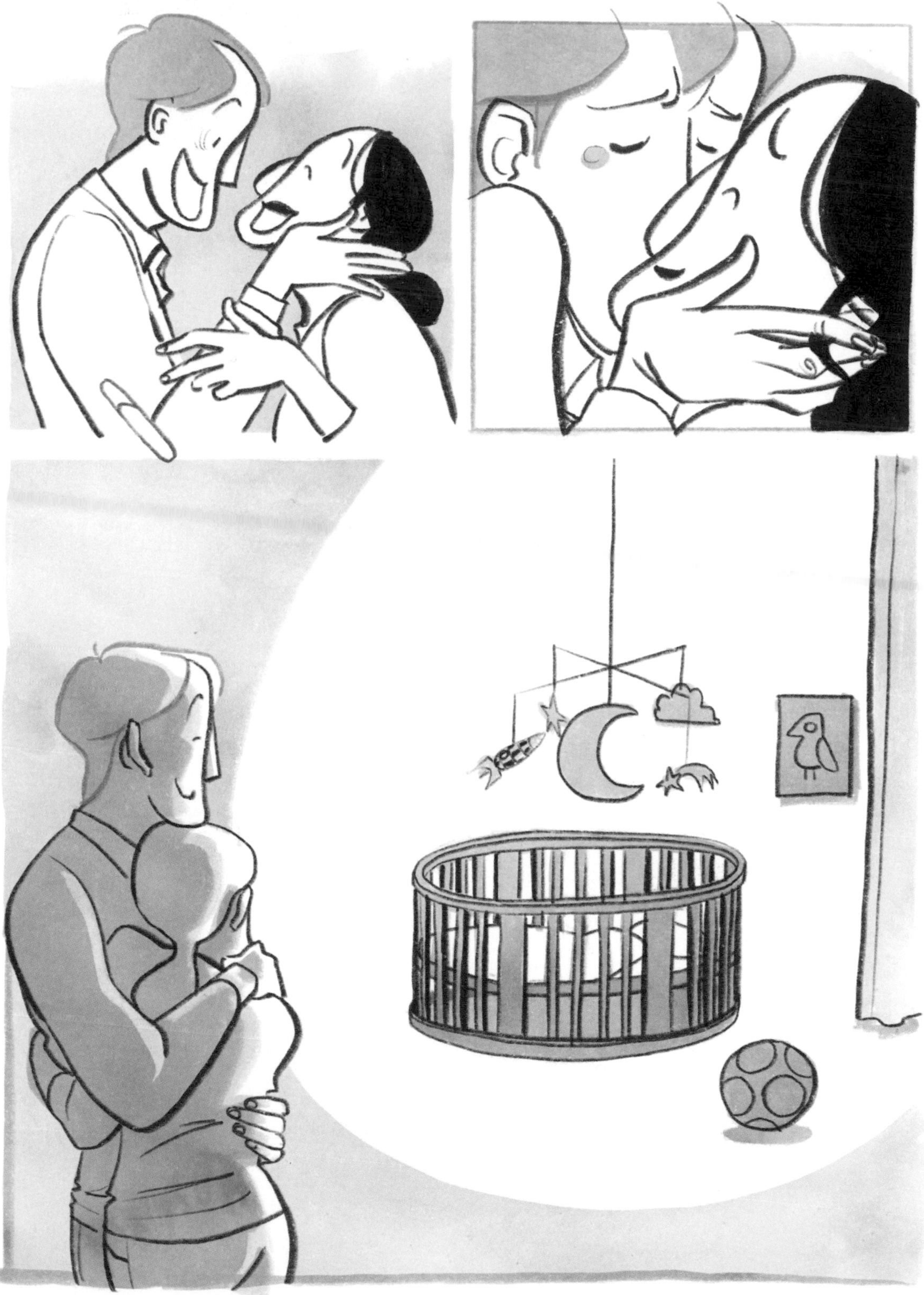

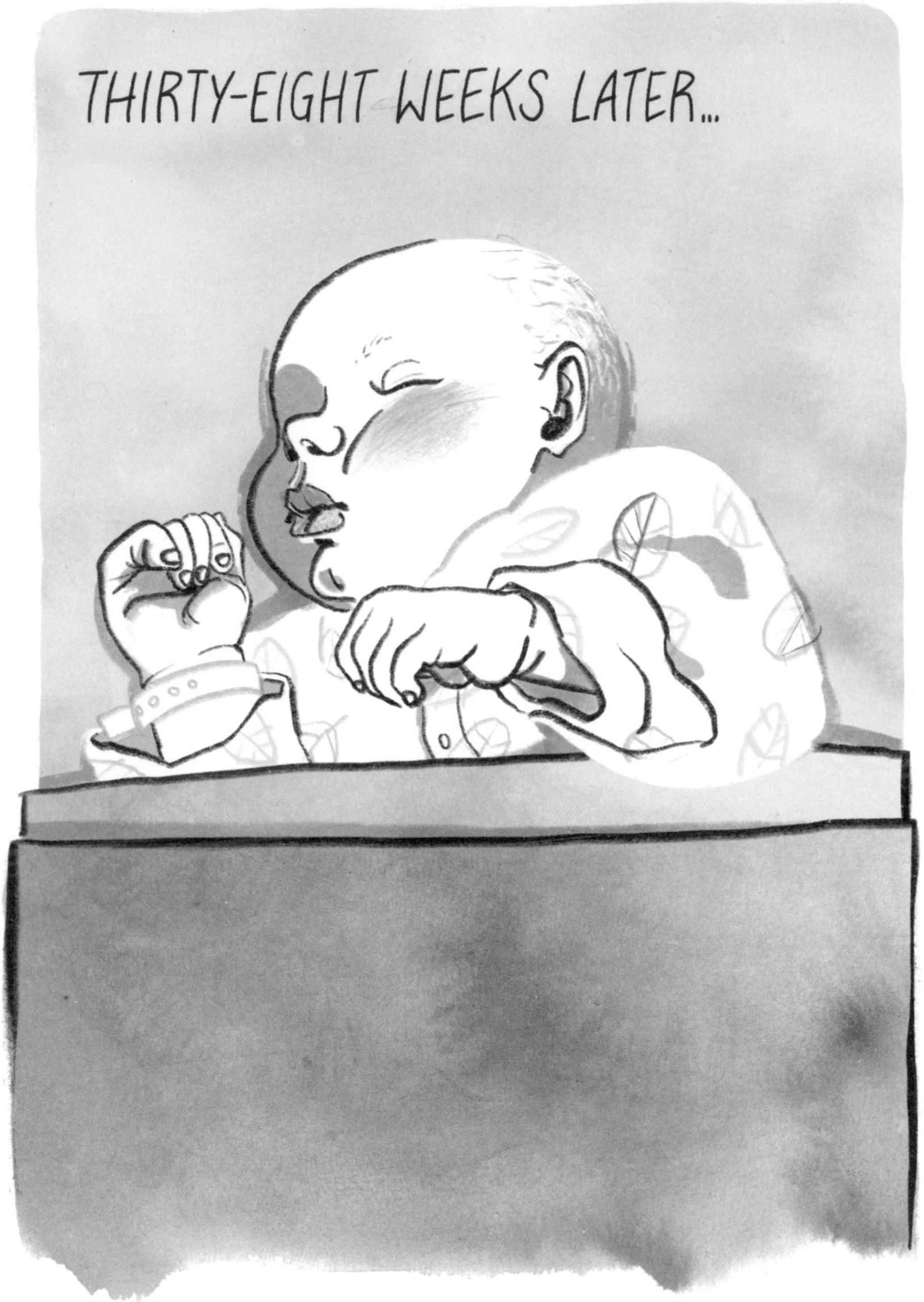
THIRTY-EIGHT WEEKS LATER...

ACKNOWLEDGEMENTS

We would like to acknowledge the honesty, generosity, and courage of the couples and individuals who agreed to be interviewed as we conducted research for this book. We hope we have done your stories justice. We would also like to thank our family for their support, and our daughters, Greta and Cleo, for their inspiration.

L.C.J. and K.J.

Thanks to the Kammer der Kritik for what it does best, the Streichelzoo for all the rest, and Maddes for the brushes.

M.W.